IRON MEN

IRON MEN

HOW ONE LONDON FACTORY POWERED THE INDUSTRIAL REVOLUTION AND SHAPED THE MODERN WORLD

DAVID WALLER

ANTHEM PRESS

Anthem Press
An imprint of Wimbledon Publishing Company
www.anthempress.com

This edition first published in UK and USA 2016
by ANTHEM PRESS
75–76 Blackfriars Road, London SE1 8HA, UK
or PO Box 9779, London SW19 7ZG, UK
and
244 Madison Ave #116, New York, NY 10016, USA

British Library Cataloguing-in-Publication Data
A catalogue record for this book is available from the British Library.

Library of Congress Cataloging-in-Publication Data
Names: Waller, David, 1962–
Title: Iron men: how one London factory powered the industrial revolution and
shaped the modern world / by David Waller.
Description: London, UK; New York, NY, USA: Anthem Press, an imprint of
Wimbledon Publishing, 2016. | Includes bibliographical references and index.
Identifiers: LCCN 2016036855 | ISBN 9781783085446 (hardback)
Subjects: LCSH: Engineers – Great Britain – Biography. | Engineering –
Great Britain – History – 19th century. | Industrial revolution – Great Britain.
Classification: LCC T55.8.W35 2016 | DDC 620.0092/241–dc23
LC record available at https://lccn.loc.gov/2016036855

ISBN-13: 978 1 78308 544 6 (Hbk)
ISBN-10: 1 78308 544 4 (Hbk)

This title is also available as an e-book.

'*The IRON MAN, as it has been called in Lancashire.*'

Andrew Ure's description of Richard Roberts's Self-Acting Mule

*

'*I'm made of iron.*'

John Thornton, Victorian industrialist, in Elizabeth Gaskell's North and South

CONTENTS

FIGURES

FOREWORD

NORMAN FOSTER

The Victorian era was certainly a heroic age of engineering, one of great foresight and vision that has not only shaped the cities we live in, but continues to serve as a blueprint for the future. When Joseph Bazalgette created London's sewer system along the Thames Embankment in the late nineteenth century, he calculated the dimensions of the pipes and doubled it – in anticipation of the capital's growth. Such was his prescience that those very pipes have adequately served the city till today. Even Stephenson's rail track and Barlow's St Pancras station have been adapted for the high-speed Eurostar that links the United Kingdom to mainland Europe, illustrating the far-sightedness of their vision – a stark contrast to the short-term sticking-plaster approach all too frequently applied to many of the infrastructure projects of today.

The ultimate expression of the age was the 1851 Great Exhibition – featuring Paxton's extraordinary Crystal Palace – that showcased the creativity displayed by British engineers in creating machines that pushed materials and technology to their limits. Such was the variety and scale of the machinery, that they literally powered Britain's transformation into the world's leading industrial nation.

The fundamental building block of all this was standardization – of machinery and parts, from the humble screw upwards, which Maudslay was instrumental in creating. Maudslay's career marks the transition from craft to industrial-scale manufacturing, from water to steam power, and wood and masonry to iron (and later steel). Without the standardized components he created, mass production would have been impossible. David Waller has done us a great favour in highlighting Maudslay's boundless creativity and energy and reveals him as the mentor and inspiration to a generation of talented engineers who changed the world. The picture of Maudslay that Waller paints is one of a man remarkably in tune with the most modern thinking. His workshop in Lambeth operated as a hothouse for creative engineering and a training ground for the best and the brightest – more akin to a Silicon Valley start-up than an oppressive Dickensian sweatshop. His exploration of materials and their properties, and his insistence on the benefits of model-making (over 40 were produced for his block-making machines!) find an echo centuries later in present practice where 3D-printing and other traditional model-making techniques are an integral part of the design process.

Reading through the book, I found many parallels between Maudslay's philosophy – '*Keep a sharp lookout upon your materials; get rid of every pound of material you can do without; put to yourself the question "[W]hat business has it to be there?",avoid complexities, and make everything as simple as possible.*' – and that of my friend and mentor Buckminster Fuller to 'do more with less'. Both of these visionary personalities exemplify an approach more appropriate than ever in an age of rapidly diminishing resources and environmental damage.

But perhaps most striking of all are the frequent references throughout the book to the sheer beauty of Maudslay's creations – a beauty born of economy of materials and pure functionality. His creations seem so perfectly fit for purpose that they take on the same qualities as most recent great works of art and design, as beautiful as a streamlined locomotive, a Chrysler Airflow, the precursor to the automobile of today, or classic and contemporary aircraft – none of these would have been possible without Maudslay's pioneering work.

Lord Norman Foster of Thames Bank is one of the world's leading architects and founder of Foster + Partners, a global studio for architecture, design and engineering.

PREFACE – THE QUEEN AND
THE MACHINES

Queen Victoria could hardly keep herself away: she visited the Great Exhibition 26 times between its opening on 1 May 1851 and early August, when she departed for Balmoral and forswore further trips to the Crystal Palace. And while she was drawn to all types of exhibits, from flowers and carpets to decorations, fine arts and furnishing, cutlery and even the American Bowie knife, she seemed to be most interested in the machinery sections. Time and time again, the Queen went back to view the machines: the new cotton machinery from Oldham and Bradford; Joseph Whitworth's machine tools, one of which was, as she wrote in her diary, 'for shearing & punching iron of just ½ an inch thick, as if it were bread!'; a knitting machine; a packing machine; a printing machine; lithographic presses; hydraulic pumps and presses; spinning and weaving machines; a curvilinear saw for timber for ships; a biscuit machine; coffee mills; 'a very curious machine for folding paper'; a machine for making combs; an immense sugar mill constructed by Robinson and Russell of Blackwall; a new kind of ship's propeller; all sorts of railway machinery and, appropriately enough, a machine for weighing sovereigns at the Bank of England.

On 11 June Victoria wrote in her diary: 'Went to the machinery part where we remained 2 hours, & which is excessively interesting and instructive, & fills one with admiration for the greatness of man's mind, which can devise & carry out such wonderful inventions, contributing to the welfare and comfort of the whole world. What used to be done by hand, & used to take months doing is now accomplished in a few instants by the most beautiful machinery.' Like so many of her subjects, she grasped immediately the implications of the Industrial Revolution. 'What a glorious, unique and truly delightful work [the Exhibition is],' she wrote in August. 'What use it has been to me in so many ways, I can hardly estimate, for it has taught me so much I never knew before, – has brought me in contact with so many clever people, I should never have known otherwise, & with so many manufacturers, whom I could have scarcely have met, unless I travelled all over the country, & visited every individual manufactury, which I should never have done.'[1]

* * *

Truly, as the early Victorian sage Thomas Carlyle declared, and Queen Victoria understood, this was the Age of Machinery.

In today's terms, mid-nineteenth-century Britain was a booming China, Asia and Silicon Valley all rolled into one, at the literal cutting machine-tool edge of every industry that mattered, from coal and steel to textiles, railways and shipbuilding. In Benjamin Disraeli's famous phrase, Britain was the workshop of the world. British industry was the undisputed technological leader, and had emerged as such, rapidly and traumatically, during the course of a few decades. For some, including the Queen and the millions who flocked to the Great Exhibition, machines heralded exciting technological change and progress. As John Stuart Mill wrote: 'The more visible fruits of scientific progress [...] the mechanical improvements, the steam engine, the railroads, carry the feeling of admiration for the modern, and disrespect for ancient times, down even to the wholly uneducated classes.'[2] The polymath Charles Babbage looked on the blast furnace of an ironworks in Leeds with a mix of awe and fear: '[T]he intensity of the fire was particularly impressive. It recalled the past, disturbed the present, and suggested the future.'[3] Carlyle himself could not help but admire the sheer physical force of the new machines, but agonized over the human costs:

> The huge demon of Mechanism smokes and thunders, panting at his great task, in all sections of English land; changing his shape like a very Proteus: and infallibly at every change of shape, oversetting whole multitudes of workmen, and as if with the waving of his shadow from afar,

1 Queen Victoria's diaries are available online at http://www.queenvictoriasjournals. org/home.do Accessed 26 June 2015.

2 J. S. Mill, 'M. de Tocqueville on Democracy in America', *Edinburgh Review*, October 1840, cited in Maxine Berg, *The Machinery Question and the Making of Political Economy* 1815–1848 (1980), 11.

3 Charles Babbage, *Passages from the Life of a Philosopher* (1864), 231.

hurling them asunder, this way and that, in their crowded march and curse of work or traffic.[4]

Amid the truly voluminous literature on the Industrial Revolution, there is much on the social impact of mechanization, but surprisingly little about the machines themselves and the men who built them. The story of precision engineering has been confined to scholarly monographs, incomprehensible to those who cannot tell their pistons from their crankshafts. Granted, there is a great deal on the heroic figures of George and Robert Stephenson or Isambard Kingdom Brunel, but while these were men of genius – men who did more than anyone to create the railway age – they were not the finest *mechanical* engineers of the age. That accolade belongs to Henry Maudslay (1771–1831), a man who rose from humble origins to become the most influential mechanical engineer of the pre-Victorian period, and to those who worked with him, absorbed his ideas, and spread his influence. In 1810, Maudslay opened a factory in Lambeth, south London, which produced the most exquisitely constructed machines and engines and became a magnet for the next generation of great engineers. Just as Google or Goldman Sachs, respectively, attract the best software or financial engineers today, Maudslay's factory became a nursery of engineering talent. Men like Richard Roberts, Joseph Clements, James Nasmyth and Joseph Whitworth learnt the art of building machines that could beget other machines, setting in motion a mechanical reproductive cycle. Within a matter of decades, machine tools had replaced hammer and chisel, and colossal ships, bridges, locomotives and engines could be built in scale and strength, and to levels of accuracy unimaginable a generation before.

Richard Roberts (1789–1864), a Welshman who, like Nasmyth and Whitworth, made his way from Maudslay's London factory to the great nineteenth-century shock city that was Manchester, built machines for use in the textiles industry. He was probably the most talented of the entire Maudslay circle, but the least successful in commercial terms. Obsessed with invention, he took out more patents than the others and is credited with multiple innovations in textiles machinery, clockmaking and shipbuilding. In the mid-1820s, he invented the self-acting mule used in cotton spinning, said at the time to be 'a machine apparently instinct with the thought, feeling and tact of the experienced workman – which was ready in its mature state to fulfill the functions of a finished spinner'.[5] This machine, hailed by Karl Marx as

---

4 From Thomas Carlyle's *Chartism* (1839), in *Selected Writings*. London: Penguin (1980), 174.
5 Andrew Ure, *The Philosophy of Manufactures*, 3rd ed. (1861), 367.

the most important of the Industrial Revolution, was so miraculously like a skilled human being at work, that it was dubbed the Iron Man by wondering contemporaries. Indeed, many of the machines made by Maudslay and his disciples had anthropomorphic qualities, seeming to function as an extension of the human mind and body, and they are collectively the Iron Men of this book's title.

Of course, the engineers themselves were men of iron, an extraordinarily talented group, who sprang largely from humble backgrounds to exercise mastery over metal, bringing unprecedented accuracy, delicacy and imagination to bear on iron and steel. These men initiated mass production and underground tunnelling, and created interchangeable components and machine tools – and one of them even built the first computer. Without the Iron Men, there would have been no Great Exhibition, no railways or steam-powered transatlantic ships. 'In the space of one lifetime mechanical engineering technique was completely revolutionized by Maudslay's example,' writes L. T. C. Rolt, the great chronicler of Victorian engineering. 'It was no coincidence that the same period saw the spectacular conquests of steam power on rails and on the sea and the complete transformation of ingenious machines for hand methods. Such dramatic developments could never have come about without the "behind-the-scenes" revolution in the engineer's workshop that was wrought by Henry Maudslay and his school.'[6]

Maudslay himself was long dead by 1851, but the company that still bore his name won a prize at the Great Exhibition, and several of the men who had worked under him had their machines on display, including Nasmyth, the inventor of the steam hammer, Richard Roberts and Whitworth, the Manchester industrialist who won more prizes than anyone else and who by the mid-century was recognized as the foremost mechanical constructor of the age.

This book tells the story of Henry Maudslay and the men who were trained in his factory; it is an account of how the early Victorians invented engineering, which is taking something of a liberty in that Henry Maudslay himself was not a Victorian (he died in 1831). But many of those featured here lived on to become the very ideal of the mutton chop–whiskered Victorian professional gentleman. The pre- and early Victorian era was a time of especial creativity, when amid tumultuous social change the British invented precision engineering, drawing their expertise and influence from an eclectic

---

6 L. T. C. Rolt, *Tools for the Job: A Short History of Machine Tools* (1965), 91.

range of sources, from instrument making to millwork, applied to a succession of apparently insurmountable practical problems. Later on, by the time of the Great Exhibition, the engineering profession had settled down, like British society as a whole, and for all the very solid achievements of the Victorian engineers, the era of the greatest creativity was past. The profession had, in short, become institutionalized.

So this account follows a broadly chronological path, starting with the age of the Regency, when London was the world's great city and the world capital of engineering. Then, perhaps more in keeping with our traditional understanding of this phase of the Industrial Revolution, the centre of gravity shifts north, to Manchester, where some of Maudslay's most talented men set up factories that grew into enormous engineering enterprises. I return to the Great Exhibition and its aftermath, when the techniques of precision engineering, hitherto deployed for exclusively peaceful purposes, were at last, and perhaps inevitably, applied to the production of ever more deadly, accurate and powerful weapons. The focus throughout is on the personalities and engineering culture associated with Henry Maudslay and his factory, an organizing principle that allows me to range broadly from machine tools to railways and weaponry. I conclude by reflecting on the implications of mechanization and Henry Maudslay's legacy. The book is thus intended as a nuts-and-bolts history of engineering enterprise in the pre- and early Victorian age, quite literally, in that the development of proper screws, nuts and bolts was one of the great technological advances of the period. Another was the dawn of standardization and mass production, described by economic historian Joel Mokyr as the 'the most underrated technological innovation of the Industrial Revolution'.[7] This is where Maudslay first made his mark.

---

7 Joel Mokyr, *The Enlightened Economy* (2009), 343. The passage is worth quoting in full. 'Standardisation of both output and input may well be the most underrated technological development of the Industrial Revolution. It required both technological and institutional breakthroughs: the sophisticated machine tools that created parts and products that were homogeneous, and the coordination on exact standards by different producers. [...] It is precisely here that the advances in machine-tool making paid their largest dividends. The screw-cutting lathes built by Henry Maudslay and his student Joseph Whitworth produced gears, screws, and bolts, of unprecedented accuracy and tolerance.'

CHAPTER 1

MAKING BLOCKS AND BORING MACHINES – THE PORTSMOUTH BLOCK FACTORY

On the morning of 14 September 1805, at the George Inn, Portsmouth, one celebrated resident rose early, scribbled a note of farewell to his mistress and began preparations for what would be his last voyage. He spent the next few hours dealing with petitions, saying official farewells and making inspections, before being rowed out to join his ship, receiving the acclaim of crowds gathered on the shore to see him off. Thus, Admiral Lord Nelson embarked on his journey to Trafalgar and immortality. The most significant event of Nelson's last morning ashore was a visit to the block-making mill in the heart of the Royal Dockyards.

Pulley blocks are unremarkable objects, but at the time of existential conflict with France, the ability to produce them in great numbers was of vital strategic importance for Great Britain, with each 74-gun warship requiring some fourteen hundred of these devices. They ranged in length from a few inches to several feet and were essential for raising and lowering the sails, masts and yards, ensuring the manoeuvrability of the vessel. 'We will apologise for our entering into so long an account of the manufacture of an article so trifling as a ship's block,' wrote the authors of the 18-page entry on the block mills in the influential Rees's *Cyclopedia*, a sort of Wikipedia for the early nineteenth century. 'Though […] this should not be despised, when its importance to naval affairs is considered, and how often the safety of a vessel may be endangered by the failure of a single block, regulating any important action in a ship's working. It is of great consequence that these […] should be made in a most accurate and substantial manner.' In 1800, the Royal Navy required more than a hundred thousand blocks a year, and they all had to be made by hand by skilled craftsmen. The work was often unreliable and inaccurate: according to one account, the pinholes in the blocks were 'so rough and so uneven that the blocks and shivers often caught fire through the violence of the friction.'[1] At a time of a labour shortage caused by the war, blocks were also expensive to make.

Once completed in 1808, the factory Nelson visited became a tourist attraction. 'It is invariably a source of delight and wonderment for visitors,' noted a newspaper account of a later royal visit. 'The princess witnessed the transformation of a cube of elm into a perfect block [...] in the space of about eight minutes, and almost by magic.'[2] The machines had not reached this peak of perfection at the time of Nelson's visit, but he too would have witnessed the operation of the world's first assembly line used for mass production, an invention arguably as important for the future of the world economy as Boulton and Watt's steam engine.

The block mills were the result of the vision and labour of three remarkable men. The first was General Sir Samuel Bentham (1757–1831), younger brother of Jeremy Bentham, the eminent philosopher and social reformer whose embalmed corpse is preserved to this day in a glass case at University College, London, the literal embodiment of Enlightenment thinking. In March 1796, Samuel was appointed as the first (and only) Inspector General of Naval Works, a role that gave the 38 year old unprecedented power over the administration of the navy, and the opportunity to put Enlightenment ideals into practice. The second was Marc Isambard Brunel (1769–1849), engineering genius and exile from revolutionary France, father of the more famous Isambard Kingdom Brunel. The third was Henry Maudslay, a young man on track to become the most influential mechanical engineer of the early nineteenth century. Together, they represent a combination of Enlightenment thinking, visionary design and practical engineering skills.

'A man of very considerable attainment and general powers, with a decided genius for mechanical art' is how John Stuart Mill described Samuel Bentham.[3] The son of a successful lawyer, Samuel was the youngest of seven children, of

---

1 Carolyn C. Cooper, 'The Portsmouth System of Manufacture', *Technology and Culture*, 25, no. 2 (April 1984): 184.

2 This was Princess Beatrice, Queen Victoria's youngest child, who sailed on the royal steamship *Alberta* from Osborne House on the Isle of Wight across the Solent to Portsmouth on 12 January 1884, along with Princess Frederica of Hanover. See *The Times*, 14 January 1884. She also saw a seven-ton Nasmyth steam hammer in action, inspected a ship's engine room and witnessed electroplating, but the block-making machinery was 'decidedly the most interesting' part of the day.

3 J. S. Mill, *Autobiography* (1873), Chapter 2.

whom only he and Jeremy survived. The younger Bentham left Westminster School at the age of 14 to train at the Woolwich shipyard, having abandoned the prospect of an academic career because of his 'uncontrollable desire to become a naval constructor'. He completed his seven-year apprenticeship, but was unenthusiastic about becoming a mere shipwright in the Royal Navy. In 1779, at the age of 23, he set off for St Petersburg and Moscow, armed with a handful of introductions and a lust for travel, excitement and self-advancement, all of which he was to find in abundance. At a time when Russia was hungry to absorb the ideas of the West, Bentham's talents as an inventor and engineer were soon recognized. The anglophile Prince Grigorii Potemkin, the favourite of Empress Catherine the Great, took him on and sanctioned a mission to the Urals to study mines and manufactories. On his return, Bentham embarked on a scandalous love affair with Countess Sophia Matushkina, the beautiful niece and ward of the governor of St Petersburg – Field Marshal Prince Alexander Golitsyn – and heiress to two fortunes. This came to the attention of the empress herself who, rather than frowning on the liaison between a scion of high nobility and a penniless English engineer, at first encouraged it, if only for the amusement of teasing the Golitsyn family. Bentham wanted to marry Sophia and, in any case, was confident that it could do no harm to his career. 'I am fully disposed that a desire Her Majesty has to assist my Match goes a great way in disposing her in my favour,' he wrote to his brother. 'She fully believes it was my love induced her to offer my Services.' Although the love match was in the end subject to an imperial veto, Bentham's path to power and prosperity opened up. Potemkin appointed him a lieutenant colonel in the Russian infantry, complete with a raffish uniform of green coat with scarlet lapels, scarlet waistcoat with gold lace, and white breeches.

In 1784, the prince sent the young Englishman to run his one-thousand-square-mile estate at Krichev in Belorussia, on the border of Poland, from which these lands had recently been annexed. Krichev was bigger than any county in England and home to around forty thousand serfs and their dependents, drawn from 50 different nationalities and living in 145 villages and 5 townships. It was also the location of what we would call a military–industrial complex, comprising mirror factories, a brandy distillery, a tannery, a copperworks, a sailmaking works, greenhouses and a shipyard. 'The estate […] furnishes all the principal naval stores in the greatest abundance by a navigable river which […] renders the transport easy to the Black Sea,' Bentham wrote home to England, awed by the scale of his new responsibilities and the opportunity for self-enrichment. He was given carte blanche to develop the industrial operations as he wished, and an unlimited budget. He established a brewery, with the aim of weaning the peasantry off vodka and instead drink wholesome English beer. He cultivated botanical gardens, put all the mills on a profitable basis, sought to set up dairies to make the finest butter and cheeses and even built a prototype of what would become known as the *panopticon*, first envisaged as a factory where low-skilled workers on the estate, a rabble of Russians, Polish Jews, Cossacks and even Geordies (the latter imported by Bentham), who knew no language in common, were put to work by the thousands under the watchful eye of management. His brother Jeremy came out to Krichev for a year to help with this project, and under his influence the concept would be developed, with far less practical success, as a form of enlightened prison for Regency England. 'I seem to be at liberty to build any kind of ship,' Samuel Bentham wrote, 'whether for War, Trade, or Pleasure.' He constructed gun frigates for the navy and barges for trade, but perhaps his most notable invention was the *vermicular*, an ingenious form of articulated riverboat, so called because it looked and moved like a worm. This vessel proved extremely comfortable and was used by Empress Catherine herself during her grand tour through the south of the empire.[4] Bentham's spell as de facto satrap of an entire province came to an abrupt conclusion in 1787, when the prince sold the estate. But this was not the end of his glittering Russian career. Appointed brigadier general and heaped with decorations, including the Order of the Cross of St George and a gold-hilted sword, he served as commander of a naval squadron during the Russo-Turkish war of 1787–92, singeing an eyebrow in a victorious encounter on the Dnieper estuary. He was given free rein to travel through the empire, ranging as far as the border with China, with an eye less to exotic scenery than to what he could see of the fur trade with Alaska or of Russian industry.

Shortly after his return to England in 1791, he visited the principal manufacturing districts, finding to his surprise that during his absence overseas 'little advance had been made towards substituting the invariable accuracy of machinery for the uncertain dexterity of the hand of man'. Steam engines were extensively employed for giving motion to pumps for raising water from mines, to machinery for working cotton, and to mills for rolling and for some other work in metal. It intrigued him that steam power was not used to propel woodworking machines, 'beyond a common turning-lathe and some saws'.[5]

---

4 Maria Bentham's *The Life of Brigadier-General Sir Samuel Bentham, KSG* (1862), 76. See also Simon Montefiore's 'Prince Potemkin and the Benthams', *History Today*, 53, no. 8 (August 2003).
5 Maria Bentham, op. cit., 97.

He had designed a machine to manufacture window sashes and was curious as to why steam technology could not be more broadly applied. It was a paradox that the practical application of steam power was still quite limited three quarters of a century after Newcomen devised his 'atmospheric' engines (so called because they were driven by atmospheric pressure) and more than a decade after Boulton and Watt's more powerful engines became generally available.

The Benthams' father died, leaving a house in the heart of Westminster, where the younger brother established a workshop. The brothers built a scale model of their hub-and-spoke panopticon, attracting a frenzy of attention when put on display. The invention never found practical application as a prison, although a century and a half later it would delight Michel Foucault, the French philosopher, who found in it a metaphor for the oppressive power of the modern state. Samuel Bentham postponed plans to return to Russia and, instead, sought at home a position commensurate with his experience and powers. In 1796, three years into the war with revolutionary France, his badgering of the Admiralty paid off, and the post of inspector general was created for him. This would not be on the same grandiose scale of his appointments in Russia, but he was given unprecedented executive power to pursue reforms on an efficient and economic basis. It was an opportunity to remake the British Navy, as vital a job as any at a time when Britain was locked in mortal combat with its ancient foe. His contract gave him explicit individual responsibility for his actions and required him to write down the basis for all his recommendations, in justified anticipation of perpetual conflict with the naval establishment. He was encouraged by the Admiralty to ride roughshod over the Naval Board, the body entrusted with naval administration, but which in the words of one contemporary observer served to 'enlarge patronage, decrease responsibility and multiply the links in the official drag-chain of the naval service.'[6] Bentham worked from first principles, beginning 'by classing the several operations requisite in the shaping and working up of materials of whatever kind, wholly disregarding the customary artificial arrangement according to trades.'[7] If machines could do the job of men, so much the better.

Just as in Russia, Bentham bustled with energy and invention, crediting himself in due course with 20 categories of innovation, from new tools to shallow docks, floating dams and mud barges. Among his more significant reforms was

6 Richard Beamish, *Memoir of Sir Marc Isambard Brunel* (1862), 52.
7 Maria Bentham, op. cit., 98.

the introduction of steam machinery into naval arsenals and dockyards. The navy had long resisted the use of steam power ashore, and would hold out against steam power afloat for years after the first engine-driven commercial boats plied the waters of the Clyde and the Thames. 'To the introduction of steam engines [...] a variety of objections were started, such as apprehensions of the derangement of established practice; doubts of the efficacy or economy of machinery as applied to naval purposes; dread of danger of fire from the use of steam engines, and of risings of the artificers at the introduction of machinery to diminish the need of their skill as well as of their labour: of these objections, and a variety of others equally groundless that were at the same time started, subsequent experience has proved the futility'. Bentham saw to it that a steam engine was erected in the Portsmouth dockyard, on the pretext that it would be used to pump water rather than power machines. 'By degrees the advantages of this primum mobile, in pumping up water, were seen and acknowledged; by degrees other and larger steam engines have been introduced, and their use has been by my means extended gradually to other purposes, as prejudices could be removed, till now at length I have the satisfaction of seeing the objection to the use of steam engines and machinery in [H]is Majesty's naval arsenals entirely done away.'[8]

The navy's sawmills had always been run along hidebound lines, sawyers entitled to traditional perquisites such as the taking of chippings, offcuts from the main operations, in the same way that ships' cooks were allowed to take the slush, a mixture of fat and salt, from the meat that they cooked (hence the expression 'slush fund'). Such practices were the enemy of efficiency. Bentham put a stop to this and made sure that every piece of timber was properly accounted for. He developed machines to help with the process of sawing logs to make planks. He also recognized the need to make pulley blocks more efficiently, and took out a patent for a machine to manufacture the sheave, the wheel that runs inside the block. To go further than this and manufacture the entire block would require the replication of two dozen carpentry tasks traditionally carried out by skilled craftsmen, from mortising, strapping and spindling to boring, scoring and broaching.[9] The carpenter would start off with

---

8 Cited by Bentham's widow, Maria Sophia, in her 'Paper on the First Introduction of Steam Engines into Naval Establishments', John Weale's *Quarterly Papers on Engineering*, vol. 6 (1846).

9 Not all blocks were made entirely by hand: the Southampton firm of Taylors had supplied the navy with blocks for generations and used basic machines, for example, circular saws, powered by water. See H. W. Dickinson, 'The Taylors of Southampton: Their Ships' Blocks, Circular Saw, and Ships' Pumps'. *TNS*, vol. 29, no. 1 (1953): 169–78.

a lump of solid elm, through which he cut as many as four oblong mortises, into each of which he would fit circular pulleys (called sheaves or shivers) made out of lignum vitae, a harder wood, which in turn rotated on a pin made of iron. The outer form of the pulley was rounded off into a smooth oval shape and then was grooved so a rope could fit round it. It was not until Bentham was introduced to the émigré engineer Marc Isambard Brunel that he began to see the outlines of a solution.

* * *

Like Bentham, Brunel made his name and career away from home. Born into a family that had farmed near Rouen for centuries, he was originally intended for the Church, but the priests entrusted with his education recognized his gift for mathematics and his intensely practical bent. He drew his tutors perfectly from life and, at the age of 11, made a barrel organ, conclusive proof that he had no religious vocation. His education was put in the hands of François Carpentier, an old friend of the family who was American Consul at Rouen and a retired sea captain, a fateful connection with the United States, which later saved Brunel's life and made his career. He was encouraged to study trigonometry and hydrography as a prelude to a career in the navy. He served from 1786 to 1792 aboard the corvette *Le Maréchal de Castries*. He impressed all with a perfectly working quadrant he fashioned out of ebony and which required a precocious knowledge of geometry, trigonometry and mathematics.

Returning to Paris from a voyage to the West Indies when the French Revolution was in full swing, his life nearly came to a premature end. Louis XVI was but days from execution, and aristocrats were being dragged from their houses and hanged from lanterns in the street. The patriotic young man was overheard in a café voicing some mildly disparaging remarks about Robespierre. And as he got up to leave, he said to his dog, "Allons, citoyen." Before the mob could turn on him for his misplaced jocularity, another unfortunate caught the attention of the crowd, giving Brunel the chance to flee. He made for Rouen. There, he laid low under the protection of the American consul and found time to meet and fall in love with Sophia Kingdom, the beautiful, orphaned 16th child of a Plymouth dockyards contractor. Sophia had rashly been sent to France to improve her language skills at precisely the wrong time. The terror gathered pace; there was fighting in the streets of Rouen; and only thanks to the intercession of Carpentier could Brunel obtain passage to New York. On the voyage, he again narrowly escaped with his life when revolutionaries stopped his ship – he presented a perfectly forged set of papers. Meanwhile, Sophia was taken captive and lived in daily peril of the guillotine.

On arrival in the United States, Brunel was appointed to survey the course of a canal to link the Hudson River with Lake Champlain. Afterwards, he returned to Manhattan where, like Bentham in St Petersburg, his skills were quickly recognized. At the age of 27, Brunel was appointed chief engineer for the City of New York, and in 1796 was honoured with American citizenship, having helped reshape New York's naval defences and built a cannon foundry. By this time, revolutionary France was at war with England, but it would still have been too dangerous to return to his homeland. At dinner in New York one night with a British general posted to America and a Frenchman just arrived from Europe, the talk turned to the spectacular English naval victories at Cape St Vincent and Camperdown, and from there to English naval superiority and the specific question of 'naval architecture and the supply of materials of ships of war'. If the rampant English had an Achilles heel, it was ships' blocks: the well-informed Frenchman told Brunel about the large and increasing cost of their production. There followed a detailed discussion of the manufacturing challenge, and that very night Brunel conceived a shaping machine.

Early in 1799, Brunel resolved to go to England, in pursuit of fame, fortune and Sophia, who by this time had been released and made her way back home. After six years of separation, she was waiting for him, and they were married on November 1 of that year. Reunited with Sophia, he would become famous, but fortune eluded him: though Brunel was one of the great engineers of the era, he was also hopelessly uncommercial. Soon after he arrived in England, he patented an ingenious machine for copying drawing and writing, a far-sighted experiment with the science of perfect replication that was at the heart of much of his mechanical experimentation. He failed, however, to take out a patent for a much more useful device, one for twisting cotton-thread into balls, which found application throughout the textile industry, but earned him no money. He also came up with a machine for making borders and hems for linen and cambric, a prototype of the sewing machine, and one for dealing and shuffling a pack of cards, neither of which found a market.

Brunel came to England armed with a letter of introduction from General Hamilton to Earl Spencer, First Lord of the Admiralty. Legend has it that he met the peer only after he presented one of his card-shuffling machines to Lady Spencer, who was bowled over by the ingenuity of the device and told her husband. Brunel continued to work on ships' blocks, taking out patents, and in

1801 submitting his designs to the company that had the contract to produce them for the navy. As a Frenchman, he was regarded with considerable suspicion – he was not allowed into the dockyards without a security escort – and he was told that the blocks had been made in the same way for 25 years, and that there was no need for change. It helped but little that he had married into an unimpeachable naval family and one of his brothers-in-law was under-secretary to the Navy Board. The following year Brunel was introduced to Bentham, who had been tinkering away amid all his other duties, but immediately recognized the superiority of Brunel's designs, acknowledging their 'great ingenuity and mechanical skill'. An arrangement was made whereby Brunel would be paid one guinea a day, together with expenses, while the machinery was being developed. In addition, he was promised a deferred and potentially enormous bonus equivalent to no less than one year's total savings made to the public once the mill was up and running. This meant that Brunel had to live on a relative shoestring for years, while all his energies were absorbed in the herculean task of bringing the machines into production.

Although Brunel's drawings were extraordinarily skilful, deploying the latest French techniques for rendering objects in three dimensions, he himself could not *make* the machinery they represented. Production would have to be outsourced to a gifted *mechanist*, a craftsman who could make the machinery with his own hands. A mechanist was a man of practical ability who shared some of the qualities of the artist, at least according to Samuel Taylor Coleridge, who wrote: 'To be a musician, an orator, a painter, a poet, an architect, or even a good mechanist, presupposes genius; to be an excellent artisan or mechanic, requires more than an average degree of talent.' A mechanic, or artisan, was thus more of a hand than a head, in the parlance of the time, although in practice many of the great engineers who feature in this book started life as humble mechanics. Great Britain had a long tradition of craftsmanship, with a high reputation for making exquisite scientific instruments, for example, and for the no less useful craft of building and maintaining water- and windmills; indeed, the word mill, used to describe the largely water-powered textile and other (manu)factories that sprang up in the late eighteenth century, reflects this heritage. Water continued to power much of British industry right until the 1830s and 1840s. Many of the first generation of great nineteenth-century engineers started their careers as millwrights, including canal builders James Brindley and John 'the Almighty' Rennie, Sir William Fairbairn, the pioneer of belt-drive systems for powering factories and builder of ships and bridges, and Sir William Cubitt, builder of canals, railways and eventually the Crystal Palace. For a long period, millwrights looked down on engineers as arrivistes.

'They thought it rather a disgrace,' and went on strike, rather than be forced
to work for mere engineers, Henry Maudslay explained to a group of MPs in
February 1824.

The Industrial Revolution was well under way, but manufacturing was
still scarcely less primitive than it had been decades before when James Watt
struggled to get his new steam engine constructed. Watt's engine was a leap
forward from the atmospheric engines constructed by Thomas Newcomen in
the early years of the eighteenth century, which were powered by atmospheric
pressure. Watt's invention was 'so much in advance of the mechanical capability of the age, that it was with the greatest difficulty it could be executed,'
wrote Samuel Smiles. 'When labouring at his invention at Glasgow, [Watt]
was baffled and thrown into despair by the clumsiness and incompetency of
his workmen.'[10] Even when Watt moved south to Birmingham, and Matthew
Boulton's best workmen were put at his disposal, no less a figure than John
Smeaton predicted that the engines could not be built. The patent for Watt's
revolutionary separate condenser was taken out in 1769, but the first engine,
about the size of a semi-detached house, was not installed until 1776. The
engines could be constructed only when the renowned Shropshire ironmaster,
John Wilkinson, developed a boring machine that produced the close-fitting
cylinders required by Watt's invention. Despite the efforts of Darby & Co in
Coalbrookdale, who produced other key components for Watt's engines, there
were no interchangeable parts. Matthew Boulton told Watt he wanted his
manufacturing operations to attain the precision of a maker of mathematical
instruments, and he liked to claim that the engines were turned out like the
trinkets produced in enormous volume at his Soho button factory, but this
was braggadocio, fine propaganda for the new product. In fact, each engine
was produced to order, components like pistons, pumps, condensers and steam
nozzles bought in from specialist manufacturers and delivered to the customer,
where an 'engine-erector' would be employed to put the machine together.

The job of looking after it was then entrusted to mechanists who lived
nearby and could be relied upon to patch it up if things went awry. It is for
this reason that William Murdoch, Watt's talented assistant, spent much of
his early career in Cornwall, rather than Birmingham, where it was his job to
tend the beam-engines that pumped water out of the tin and copper mines that
were Boulton and Watt's principal early customers. While still in Cornwall, he
invented gas lighting, which in due course transformed the factory economy

---

10 Samuel Smiles, *Industrial Biography: Ironworkers and Toolmakers* (1863), 179.

as machines could be kept running 24 hours a day. He also patented a steam carriage, a remarkable precursor of the next century's revolution in transport, and was responsible for the sun and planet gear system, which when patented in 1781 triggered the mass use of Watt's steam engines in textiles factories, as at last a way was found to transfer the power from the up and down motion of the piston to the circular motion required to impel machinery.[11] By the turn of the century, there were more than 2,000 steam engines installed in Great Britain, more than a quarter manufactured by Boulton and Watt. Still, even at the end of the Napoleonic Wars, virtually all machinery and mill equipment was made by hand, on the spot, in the mills and factories where it was to be used.

In Manchester – where the textile industry generated tremendous demand for shafting and gearing equipment, waterwheels, steam engines and boilers – machinery had barely evolved from the hybrid of cast iron, brass, wood and leather that characterized the efforts of the late eighteenth century. 'There were neither planing, slotting nor shaping machines,' recalled Sir William Fairbairn, 'and with the exception of very imperfect lathes and a few drills the preparatory operations of construction were effected entirely at the hands of the workmen.'[12] As late as 1818, when Fairbairn opened his own engineering works to produce the shafting and machinery used to power cotton mills, matters were still so primitive that the two or three lathes in his possession were powered not by steam, but by an Irishman called Murphy and two of his compatriots. Many of the early mechanics were craftsmen of great skill, but the level of accuracy in their work was poor by later standards. Every machine and every engine was produced as a one-off. 'Everything depended on the dexterity of hand and correctness of eye of the workmen,' wrote Samuel Smiles. 'The work turned out was of very unequal merit besides being exceedingly costly. Even in the construction of comparatively simple machines the expense was so great as to present a formidable obstacle to their introduction and extensive use.'[13]

Machines in use at this time looked impressive, but were not in fact very powerful, as confirmed by James Nasmyth's painterly descriptions of the Carron

---

11 Rival engineers Matthew Wasborough and James Pickard had already patented the use of the more obvious crank, forcing Watt and Murdoch to find an alternative.

12 Presidential address to the British Association at Manchester, 1861.

13 Smiles, *Industrial Biography*, 212. Both Smiles and Fairbairn were exaggerating how backward Manchester had been, as the city was already a prominent centre of machinery production, home to well-known firms such as Peel, Williams & Co., and Bateman and Sherratt. It took the Maudslay men to make Manchester pre-eminent in engineering.

Ironworks at Falkirk, in Stirlingshire, then one of the largest iron foundries in the world, celebrated for forging many of the cannon used to defeat Napoleon. As a boy, he viewed the production of pig and sow iron (so-called because the molten iron was directed from one large pool to many smaller ones, like a sow feeding its young) and sketched the working engines. 'When seen partially lit up by the glowing masses of white-hot iron, with only the rays of bright sunshine gleaming through a few holes in the roof, and the dark, black, smoky vaults in which the cumbrous machinery was heard rumbling away in the distance – while the moving parts were dimly seen through the murky atmosphere, mixed with the sound of escaping steam and rushes of water; with the half-naked men darting about with masses of red hot iron and ladles full of molten cast-iron – it left a powerful impression on the mind.' But, he noted: '[They] existed at an early period in the history of British iron manufacture,' he wrote. 'Much of the machinery continued to be of wood. Although effective in a general way it was monstrously cumbrous. It gave the idea of vast power and capability of resistance, while it was far from being so in reality.'[14]

Screws, nuts and bolts, the most basic components for a machine, did not become standardized until much later in the nineteenth century. 'Every bolt and nut was thus a speciality in itself, and neither possessed nor admitted any community with its neighbours,' observed James Nasmyth. 'All bolts and their corresponding nuts had to be specially marked as belonging to each other. Any intermixture that occurred between them led to endless trouble and expense, as well as inefficiency and confusion, – especially when parts of complex machines had to be taken to pieces for repairs.'[15] It was conceivable that Brunel's vision would be too far ahead of its time, incapable of being made under prevailing conditions.

Enter Henry Maudslay, the third figure without whom the block-mill would never have come into being. His father, a sergeant wheelwright in the Royal Artillery, had been severely injured in the West Indies. A musket ball hit his throat, and he was saved only because his stock, or leather collar, took the force of the shot. Maudslay junior kept that stock and jested that, but for this item of clothing, he would not have come into existence. The invalid father got a job as a storekeeper at the Woolwich Arsenal, and Maudslay was born on August 22, 1771, in a modest two-room wooden house nearby, behind the Salutation Inn in Beresford Square. The fifth of nine children, Maudslay grew up at the heart of one of the world's great centres of naval and military power, the home

---

14 James Nasmyth, *Autobiography* (1883), 212.

15 Nasmyth, *Autobiography*, 128.

of thousands of sailors, soldiers and their families, a bustling, clangorous, smelly place where arms were made and ships constructed, and where animals were slaughtered and salted to provision soldiers and sailors overseas. When he started work at the Arsenal at the age of 12, Great Britain was at the start of what would be a more than 20-year struggle with France. Although Maudslay never fought for his country, his working life must be seen in the context of the long war. As is so often the case, war would force the pace of technological change and create opportunities for talented men. Maudslay's first client as an independent engineer was the Royal Navy and, having won the trust of the authorities, the government and armed forces would be major customers throughout his career.

Maudslay's first job was as a powder monkey, filling cartridges with gunpowder, before being promoted to a job in carpentry and, at the age of 14, to the smithy, where he rapidly developed a reputation as a craftsman of the highest skill. He knocked up highly wrought trivets in his spare time and used lead bars to model his forgings, the malleability of cold lead being similar to that of red-hot iron. Like James Watt, who began his career as a maker of scientific instruments, Maudslay was an artisan before he became an engineer but, unlike Watt, he did not have the benefit of formal schooling beyond his early teens: he acquired his knowledge on the job, by word of mouth, and by direct learning from those at work around him. At the age of 18, Maudslay came to the attention of Joseph Bramah (1749–1814), one of the great inventor-engineers of the eighteenth century, with 18 patents to his name between 1778 and 1812. Bramah was the inventor of the modern lavatory system, his ball-and-cistern water-flushing convenience becoming a staple in middle-class homes of the nineteenth century and eventually Osborne House, the royal seat on the Isle of Wight. He applied his knowledge of the motion of fluids to other vital fields of human activity, inventing both the beer pump and the hydraulic press, the latter a machine to this day with many industrial applications.

Bramah had taken out a patent for a lock in 1784, but was not able to construct and sell these useful and beautifully crafted items in sufficient scale to make them a commercial proposition – they had to be made as one-offs. On joining Bramah's factory in Denmark Street, St Giles (near Covent Garden) in 1790, Maudslay spent his first 12 months building the machines used to manufacture the lock. They were housed in a secret workshop and displayed 'a systematic perfection of workmanship, which was at that time unknown in similar mechanical arts,' according to one Victorian engineer.[16] They were

---

16 John Farey, quoted in the *Minutes of the Proceedings of the Institution of Civil Engineers*, April 9, 1850.

intricate, miniature machines, made out of metal and wood, that cut the grooves in the barrel of the lock and the corresponding notches in the keys and sliders. Micrometer screws were aligned 'so as to ensure that that the notches in each key should tally with the unlocking notches of the sliders[; …] without the machines, the locks could not have been in any great number, with the requisite precision, as an article of trade'. They were highly original, going far beyond for example the clock-maker's wheel-cutting machines. One surviving machine can be seen in the storerooms of the Science Museum. Bramah attributed the success of his locks 'to the use of these machines, the invention of which had cost him more study than that of the locks [themselves].'[17]

Maudslay stayed with Bramah until marrying Sarah Tindale, his master's housekeeper, and had the temerity to ask for an increase in his 30s a week pay, which was refused. He left Bramah's employ in 1797, when he was 26, setting up on his own account, establishing a machine shop and smithy at 64 Wells Street, just off Oxford Street in London's West End. This was barely half a mile away from St Giles. At the time, London was Great Britain's centre of specialist engineering and other light industry – for example brewing, dye works, calico printers, flour mills, tanneries, distilleries, foundries and machine makers – home to hundreds of small-scale workshops. Between 1776 and 1800, the biggest market for Boulton and Watt's engines apart from Cornwall and Lancashire was London, the first being installed in 1786 at the Albion Mills just south of Blackfriars Bridge in Southwark, to the wonderment of the population. When the mill burnt down in 1791, arson was suspected, and the blackened ruin is said to have inspired William Blake's celebrated image of 'dark satanic mills'. To give a further sense of the relative importance of London as an industrial centre, in 1804–05, when the engineer John Farey visited every steam-powered establishment in London, he counted 112 steam engines at work producing 1,355 horse-power, compared to just 32 in Manchester, which produced 430 HP. Even in 1825, there were about 290 steam engines in London, compared to 240 in Manchester, 130 in Leeds and 80 to 90 in Glasgow.[18]

One day, M. de Becquancourt, a friend of Brunel and fellow exile from France, happened to look in Maudslay's shop window. He was impressed and told Brunel about what he had seen. Brunel himself went along to Wells Street, showing Maudslay a partial drawing for one of the block machines, without disclosing what it was for. Brunel returned with another drawing, still cagey

17 Cited in H. W. Dickinson, 'Joseph Bramah and His Inventions', *TNS*, vol. 22, no. 1 (1942): 169–86.
18 Cited in G. von Tunzelmann, *Steam Power and British Industrialisation to 1860* (1978), 31.

about his true requirements. On the third visit, Maudslay exclaimed: 'Ah! Now I see what you are thinking of; you want machinery for making blocks.' Impressed by Maudslay's rare understanding, Brunel took the young man fully into his confidence, and in 1800 hired Maudslay to build scale models of the machinery. In total, Maudslay built 44 of these models, and they were eventually displayed to admiring crowds at the Admiralty Museum in Somerset House in the Strand. (They are now hidden in the collection of the National Maritime Museum in Greenwich). Bentham and the Navy Board awarded Maudslay the contract to build the full-scale machinery, a task that took the best part of eight years. He moved to larger premises in Margaret Street, Cavendish Square, in 1802. It was only in 1806 that Maudslay dispensed with muscle power and installed in his cellar a 4-hp engine linked by a belt-drive to the machines above, a typical arrangement in the factories and workshops of the day.

Fashioned out of cast iron and hardened steel, the frames decorated with Tuscan columns and pediments, the block machines set new standards for elegance as well as ingenuity and durability. In 1823, James Watt, the son of the great inventor, visited Portsmouth and praised the workmanship and appearance of the machines and wondered why Boulton and Watt could not 'adopt the same beautiful Gothic in our framing and the same neatness of finish'. The authors of the *Cyclopedia* agreed, noting that the machines 'presented an elegant proportion in their form, which [was] very agreeable to the eye'. They were fully up and running by late 1807 and still going strong more than half a century later when Samuel Smiles published his *Industrial Biography*:

> We despair of being able to give any adequate description in words of the intricate arrangements and mode of action of the block-making machinery. Suffice it to say, that the machinery was of the most beautiful manufacture and finish, and even at this day [1863] will bear comparison with the most perfect machines which can be turned out with all the improved appliances of modern tools.[19]

With cross-bracing derived from carpentry and lattice-work from the smithy, as well as the classically Georgian look to their frames, these were truly eclectic, startlingly original creations that married functionality with elegance. Each nut and matching bolt was hand-made and given a number to make sure they were correctly placed, and the movement of the cutting tools

---

19 Smiles, *Industrial Biography*, 221.

took place between carefully positioned 'stops', so that the workman did not have to use too much skill and judgment, and the risk of cutting too far or too deep was limited. Powered by a single 32-hp steam engine, the Portsmouth plant had 45 machines of 22 different types, laid out as a production line, each stage of the work flowing naturally on to the next task. Human beings had to carry the partially completed blocks from one stage to the next, but still the machines replaced the labour of 110 skilled workers with just half a dozen unskilled men and boys, thus ruthlessly achieving the objective of reducing the manpower requirements as well as the costs of production. Working faster and more accurately than mere humans, the machines produced 130,000 blocks a year, at a value of £540,000. They were the world's first set of machines built entirely of metal, 'rigidly framed in cast iron, in which accurately fitting iron shafts, wheels, gears, and feed screws were moved by steam power to apply sharp steel saws, gouges, chisels, and drill bits to the wooden blocks and sheaves'.[20] They exploited the latest in advances in iron smelting and ironworking, which 'made high quality cast and rolled iron and crucible steel more easily available for constructing machinery.' They represent a crucial link between the craft economy, and the age of the machine: Maudslay had to make the components for the machines by hand, or with the help of his self-acting lathe, but once the machines were built, they – or other machines built on similar principles – could be used to build the next generation of machines. As Nasmyth noted, looking back from the vantage point of mid-century when England's mechanical supremacy was unchallenged, they 'contained the prototypes of nearly all the modern engineer tools that have given us so complete mastery over materials, and done so much for the age we live in'.[21]

The block factory also illustrated the hugely significant difference between making and manufacturing: as Charles Babbage was later to explain, *making* implied creating a limited number of objects, while *manufacturing* meant producing lots of more-or-less identical items. The quest to make the perfect copy was something of a Holy Grail for engineers at the time: both Watt and Marc Brunel experimented with copying machines, and the development of mechanized mass production was a logical consequence of this early work. There were successful manufactories before the block factory, for example in the so-called Birmingham toy trade, where hundreds of small workshops produced tiny items like buttons, buckles, pins and nails, or the Sheffield

20 Carolyn Cooper, op. cit., 197.
21 Nasmyth, *Autobiography*, 131.

cutlery industry. Another was Josiah Wedgwood's pottery empire. But in these eighteenth-century operations, the mass production was carried out almost completely by hand, without automation and mechanization.

The pioneering practices in Portsmouth implied a radical new organization of time, space and materials. As Babbage put it: '[I]f the maker of an article wish to become a manufacturer [...] he must attend to other principles besides those mechanical ones on which the successful execution of his work depends; and he must carefully arrange the whole system of his factory in such a manner, that the article he sells to the public may be produced at as small a cost as possible'.[22] The manufacturing engineer of the future would have to know how to lay out his factory to machinery to ensure the optimal flow of materials and machinery; he would seek economies of scale from producing items in high volume; he would have to have rigorous control of costs and understanding of the time taken for each part of the manufacturing process. Bentham and Brunel both understood the power of the idea, and tried to put it into practice in a variety of ventures, but they failed to build commercially viable operations. Some economic historians refer to the 'American System of Manufacturing', as if the first mass production was developed by US engineers. This is not accurate, in that Maudslay and his disciples built some of the first factories to put these principles successfully into practice.[23]

'The block-making machines were in many ways a sign of the industrial future,' writes Joel Mokyr. 'In their close coordination and fine division of labour they resembled a modern mass-production process, in which a strongly interdependent labour force of ten workers produced a larger and more homogeneous output than the traditional technique that had employed more than ten times as many.'[24] The annual saving was an immense £16,621-8-10d. The project must count as among the most profitable partnerships between government and the private sector, as the initial outlay of £54,000 was earned

---

22 Charles Babbage, *On the Economy of Machinery and Manufactures* (1846), 81.
23 Eli Whitney (1765–1825), the inventor of the cotton gin, is often credited with developing mass production for his Connecticut musket factory in the early nineteenth century. He introduced interchangeability of components, but American industry at large did not catch on until the 1850s. There were other antecedents, for example, Matthew Boulton's Soho Mint, which produced hundreds of millions of coins and medals between the 1790s and its closure in the 1850s. See chapters 10 and 13 for Joseph Whitworth's contribution to the development of mass production, and for his reflections on the state of American industry in the 1850s.
24 Mokyr, op. cit., 344.

back in less than three years of production. After exasperating delays and humiliating petitions, Brunel was paid just over £17,000. Later, he established a veneering plant and a sawmill in Battersea, where his circular saws (between 10 and 18 feet in diameter) cut timber into wafer-thin pieces. 'His system saves both time and money and it would be impossible to imagine articles made with greater precision and utility than those produced here,' wrote a deeply impressed visiting Swiss industrialist.[25] But the sawmill burnt down, and a promising enterprise to mass-produce boots for footsore British soldiers failed after the inconvenient termination of hostilities with the French after the Battle of Waterloo. In 1821, he was declared bankrupt. Maudslay himself received £12,000, providing the capital for the expansion of his business to new premises on Westminster Bridge Road, Lambeth Marsh. The two men would later cooperate on another of the great projects of the age, the plan to build the first tunnel under the Thames. Bentham himself derived no direct pecuniary benefit and eventually found that he had made too many enemies in his pursuit of naval reform. He was dismissed from his post in 1812, albeit with a generous pension, and eventually retired to France, where he wrote his memoirs and tinkered with further inventions. He acknowledged, somewhat to his surprise, that of all his innovations, the block mills exercised the most powerful grip on the public imagination. Remarkably, the last of the machines ceased working in their original setting only in December 1968, some 160 years after they were first installed.

Some of Maudslay's boring machines are to be found in the Science Museum, where the milling, mortising, gouging, stamping and other machines sit, hugger-mugger and Heath-Robinson-esque, in the shadow of Robert Stephenson's 1829 locomotive, *Rocket*. Of the eight machines on display, the largest is the mortising machine, its flywheel 6.5 feet in diameter, which was the biggest metal machine tool when it was built, and remained in use until three years after World War II (Figure 1). The original purpose of these machines is impossible to discern in the absence of movement and noise and, despite the best intentions of the museum, it is difficult to capture a flavour of the drama and excitement felt by contemporaries when they saw the machinery in operation. The Portsmouth factory, just across the basin from HMS *Victory*, has been left semi-derelict, with no plans for refurbishment.

In the Age of Sail and of Jane Austen and the Romantic poets, however, a visit to Portsmouth gave an intimation of the industrialized future. As early as 1805, before the plant was even finished, Brunel complained that the number

---

25 W. O. Henderson, *J. C. Fischer and His Diary of Industrial England 1814–1851* (1966), 29–30.

FIGURE 1 Henry Maudslay's mortising machine, used in making ships' blocks (Courtesy of Science Museum, London)

of sightseers got in the way of the operations: '[T]his frequent admission of visitors is of great hindrance to the men at work[; …] the place was the whole morning crowded with visitors, much to the annoyance of the service.'[26] He

26 Cited in Cooper, op. cit., 213.

asked Bentham to erect a fence to keep off the pesky tourists, but that did not stop them coming, including novelists Maria Edgeworth and Walter Scott. A month after Napoleon abdicated and was exiled to the Isle of Elba, the Prince Regent invited his allies from Russia and Prussia to a celebration-cum-summit in Portsmouth. Emperor Alexander of Russia and the king of Prussia hugely enjoyed the military parades, the incessant gun salutes, the voyage out to sea to review 30 ships of the line drawn up in the Roads – but the highlight of the day was elsewhere. 'The Emperor and the King appeared more particularly interested in that unequalled system of making the ships blocks, the rapid operations of which they witnessed with particular pleasure,' reported *The Times*. 'All expressed their admiration of the mechanism, which they thought was of itself worth coming to Portsmouth to see.'[27]

---

27 *The Times*, 27 June 1814.

CHAPTER 2

MAUDSLAY'S – THE MOST COMPLETE FACTORY IN THE KINGDOM

In 1810, Henry Maudslay moved from the West End to the premises of a disused riding school on Westminster Bridge Road, Lambeth, on what is now the site of Lambeth North tube station, to the east of St Thomas's Hospital. Buoyed by his growing reputation and the orders that came with it, he sought to expand. Over the course of the next two decades he built an engineering works that endured until 1900, employing more than a thousand men by mid-century. In the mid-1820s, it was recognized as 'the most complete [factory] in the kingdom'.[1] Lambeth was home to a variety of industries, including glassworks, potteries (Doulton was based nearby), breweries and engineering works; it was also undergoing the explosive population growth more usually associated with the industrial cities of the North, the number of people living here trebling between 1808–31. Adjacent to the works in Oakley Street was the Orphan's Asylum, home for some of the overspill humanity, and the densely populated Lambeth Walk was minutes away. On the same street, but closer to Westminster Bridge, was Philip Astley's Royal Amphitheatre, the first great circus-cum-theatre, famous for its equestrian shows, and not too far away in Kennington were the Vauxhall Pleasure Gardens. The front erecting shop of the factory had an elegant classical façade, to the left of which were a block of offices and houses, one of which was occupied by Maudslay himself. Behind were forges, smithies and workshops and a bigger erecting shop measuring 148 feet by 55 feet, with walls 20 feet high, where cranes moved massive iron components to assemble engines and machines.

---

1 Thomas Allen, *The Parish of Lambeth* (1826), 319. 'Steam engines, tanks for shipping, and all works for various factories, are here executed in the best manner'. At this time, the firm employed some 200 men.

Today, the site is a nondescript area to the south of Waterloo Station, a terminus that of course did not exist when Maudslay chose to locate his business there, before the Battle of Waterloo and long before the railways. There is nothing left to remind us of Maudslay's presence, except a memorial tablet erected high on the wall inside the ticket office of the tube station, which you would hardly notice if you did not come looking for it:

> On this site between 1810 and 1900 stood the Works of Maudslay, Sons & Field famous for marine and general engineering and as the training place of many engineers of renown.

This ought to be hallowed ground for all engineers and aficionados of the Industrial Revolution, as it was for knowledgeable contemporaries.

One such was the young James Nasmyth (1808–1890), who in May 1829 sailed down to London from Edinburgh with his father, Alexander, and his brother, George, looking for a job. Nasmyth would become one of the most successful and prosperous engineers of the mid-Victorian era. At the time of his first visit to London, he was a 20-year-old self-taught engineering obsessive who spent his spare time visiting foundries and examining the workings of steam engines. The best engineers of the day included Matthew Murray of the Round Foundry in Leeds, James Fox of Derby, and Bryan Donkin and John Penn of London. But, wherever Nasmyth went to look at an engine, above all others he heard the name and fame of Henry Maudslay. 'I was told that his works were the very centre and climax of all that was excellent in mechanical workmanship,' Nasmyth recalled. He wanted to get a sight of the factory in Lambeth, and ideally to get a job there, as a mere labourer if necessary. The modern parallel is perhaps with a computer-obsessed teenager taking a plane to Silicon Valley in the hope of getting a job at Apple or Facebook, laptop full of ingenious computer programs with which to impress his would-be employer. The Nasmyths sailed to London from Leith on 19 May 1829, taking with them a small high-pressure engine the brothers had made at home and a portfolio of mechanical drawings. After a day's sightseeing, they went to the engineer's house next to the works in Westminster Bridge Road. Nasmyth senior and Maudslay had been introduced by a mutual friend some years before, and this was enough of a connection to present themselves at Maudslay's front door.

Maudslay was happy to receive his Scottish guests. They sat down in his study and Alexander Nasmyth explained his business. Maudslay was very large and portly, 6 foot 2 inches tall, with a commanding presence, a frank and easy manner, a broad, open face, and a good sense of humour. He had the habit of taking snuff as he walked around the factory, the dust collecting in the folds of his waistcoat. 'A very kind and humane man,' is how *The Times* described him in May 1826, 'extremely attentive to the needs of the comforts of those in his employment'.[2] Maudslay said courteously that he was no longer taking on paying apprentices, as he had found the experience to be unsatisfactory. Nasmyth's heart sank. But would they like to visit the factory? Just as if our teenage programmer were to be shown round Facebook HQ by Mark Zuckerberg himself, Nasmyth could barely contain his excitement. 'The sight of the workshop astonished me,' Nasmyth recalled. 'They excelled all that I had anticipated. The beautiful machine tools, the silent smooth whirl of the machinery, the active movements of the men, the excellent quality of the work in progress, and the admirable order and management that pervaded the whole establishment, rendered me more tremblingly anxious than ever to obtain some employment there, in however humble a capacity.' Nasmyth spotted a man cleaning out the ash beneath a boiler. 'On the spur of the moment I said to Mr. Maudsley [sic], "if you will only permit me to do such a job as that in your service, I should consider myself most fortunate!" I shall never forget the keen but kindly look that he gave me. "Ah, so," said he, "you are one of that sort, are you?" Maudslay said.'[3]

Maudslay invited the young man to come back the next day to meet him and Joshua Field (1786–1863), the son of a City corn and seed merchant who had joined Maudslay when he was still in the West End. A former apprentice at the Portsmouth dockyards, Field had worked with Maudslay on the block-making machinery and moved with him from the West End to Southwark. In 1812, Maudslay took Field, together with his eldest son, Thomas, into partnership and later the firm's name changed to Maudslay, Son & Field.[4] Field would long outlive Maudslay and became a leading engineer of the Victorian era. Nasmyth turned up with his drawings and the home-forged working model of a steam engine. Maudslay and Field spent 20 minutes inspecting the work, and then took him on. 'This is where I wish you to work, beside me, as my assistant workman,' Maudslay is supposed to have said, emerging from his private workshop into the library where the young man was waiting on tenterhooks. 'From what I have seen there is no need of an apprenticeship in your case.'

---

2 *The Times*, 25 May 1826.

3 Chapters VII and VIII of Nasmyth's *Autobiography* contain his account of his time at Maudslay's, cited here and below.

4 The name changed again, to Maudslay, Sons & Field, in 1831, shortly before Henry Maudslay died, when he took two other sons, John and Henry, into the partnership.

By the time James Nasmyth set down the story of this encounter, his brother, George, had suffered a professional disgrace and had died in exile overseas. So George was airbrushed out of the account, as if he had not been there, but he too was offered a job. Nasmyth was shown around Maudslay's personal workshop, the inner sanctum of engineering creativity:

> [Maudslay] proceeded to show me the collection of exquisite tools. [...] They mostly bore the impress of his own clear-headedness and commonsense. They were very simple, and quite free from traditional forms and arrangements. At the same time they were perfect for the special purposes for which they had been designed[; ...] every tool had a purpose. It had been invented for some special reason. Sometimes it struck the keynote, as it were, to many of the important contrivances which enable man to obtain a complete mastery over materials[;...] there were hung upon the walls [...] many treasured relics of the first embodiments of his constructive genius. There were many models explaining, step by step, the gradual progress of his teeming inventions and contrivances[; ...] they were kept as relics of his progress towards mechanical perfection.[5]

Early in his career, Maudslay had developed the slide-rest, a mechanism for holding and guiding the cutting tool. Before this invention, lathes were made largely of wood, and the workman had to guide the cutting tool by muscular force alone, the finish depending on his skill and dexterity. Now the tool could be clamped into place in the rest and moved smoothly along the object being worked on, resulting in a vastly superior and more accurate end product. It was, said Nasmyth, the most powerful agent 'towards the attainment of mechanical perfection'; he drew an illustration showing one workman nonchalantly operating a slide rest, his hand in his pocket, while another struggles to get the job done in the old-fashioned way. Maudslay did not invent the slide rest, in that similar tools were deployed in the watch-making industry, and there were isolated examples of talented engineers having a similar idea, but he was the first person to bring it to a peak of perfection, and to make it part of the engineering mainstream, as Karl Marx recognized:

> It was necessary to produce the geometrically accurate straight lines, planes, circles, cylinders, cones, and spheres, required in the detail parts of the machines. This problem Henry Maudslay solved in the first decade

5 Nasmyth, *Autobiography*.

of this century by the invention of the slide rest, a tool that was soon made automatic, and in a modified form was applied to other constructive machines besides the lathe, for which it was originally intended. This mechanical appliance replaces, not some particular tool, but the hand itself, which produces a given form by holding and guiding the cutting tool along the iron or other material of labour. Thus it became possible to produce the geometrical shapes of the individual parts of machinery 'with a degree of ease, accuracy, and speed, that no accumulated experience of the hand of the most skilled workman could give'.[6]

Half a century later, Nasmyth and Samuel Smiles set down the principles on which Maudslay operated, and which came to influence nineteenth-century mechanical engineering as a whole. First and foremost, Maudslay worked with his hands. 'The truth is that the eyes and the fingers – the bare fingers – are the two principal inlets to practical instruction,' Nasmyth wrote. 'They are the chief sources of trustworthy knowledge in all the materials and operations which the engineer has to deal with. No book knowledge can avail for that purpose. The nature and properties of the materials must come through in the finger ends.' 'It was a pleasure to see him handle a tool of any kind,' one of Maudslay's old workmen told Smiles, 'but he was quite splendid with an eighteen inch file.' He exuded a calm authority, an absolute concentration on the task, which powerfully affected those who saw him work. 'Every stroke of the hammer, chisel or file, told as an effective step towards the intended result. It was a never-to-be forgotten practical lesson in workmanship, in the most exalted sense of the term.'

Maudslay was wholly absorbed in manual work of the highest order, and his skill and concentration transmitted itself to others. As he grew rich and successful, there was no question of his detaching himself from the workshop, and he remained literally a hands-on manager, living alongside the factory, involved in every aspect of the production process. On Sunday mornings, when the factory was still, he used to walk around to inspect the work in progress. 'On these occasions he always carried with him a bit of chalk which in a neat and very legible hand he would record his remarks in the most pithy and sometimes sarcastic terms,' recalled Nasmyth. His men listened intently to every terse word, and it would soon be known all over the shop who had received a memorandum in chalk from the master. Thus, through force of

---

6 Karl Marx, *Capital*, vol. I. See Chapter 15, Machinery and Large-Scale Machinery, 507 in the Penguin edition.

FIGURE 2 Henry Maudslay in his prime: lithograph portrait on stone by Henri Grevedon, 1827 (Courtesy of Science Museum, London)

H. MAUDSLAY.

personality and his own practical skill, did Maudslay regulate his workforce (Figure 2).

There is a paradox: Maudslay was a craftsman who devoted his skill to devising the very machines that in time did much to make the craftsman's role redundant. Unlike Nasmyth, however, who in his maturity was motivated by a strongly commercial desire to reduce skilled labour and, therefore costs, in his own and customers' factories Maudslay saw mechanization as a way of empowering his men, not eliminating them. His first lathes and turning-machines enabled the skilled operator to transcend his own limitations of eye and muscle. His machines brought the accuracy and strength required for operating in metal rather than wood, and for making other machines of scale rather than the miniaturist locks where he had learnt his trade. His quest was mechanical perfection, rather than financial reward. 'In designing and executing [his work],' wrote Samuel Smiles, 'his main object was to do it in the best possible style and finish, altogether irrespective of the probable pecuniary reward. This he regarded in the light of a duty he could not and would not evade, independent of its being a good investment for securing a future reputation.'[7] Maudslay's philosophy was elegantly minimalist:

> First, get a clear notion of what you desire to accomplish, and then in all probability you will succeed in doing it. [Then:] 'Keep a sharp look out upon your materials; get rid of every pound of material you can do without; put to yourself the question, 'what business has it to be there? Avoid complexities, and make everything as simple as possible.

He had the rare ability to think through mechanical interrelationships in three dimensions, and to body forth those ideas in metal: it was said that he could see the finished forging in a lump of red hot iron, just like a sculptor can visualize the final statue in a block of marble. His inventions were things of beauty, as indeed one can still see from the block-making machinery on display at the Science Museum; but the elegance and simplicity of the design was part of their absolute functionality, a complete fitness to purpose. He abhorred sharp angles, preferring the corners of his machines to be rounded off. This was both natural, imitating the way fingers are joined to a hand or branches to a tree, and practical: his machines were stronger that way. ('Sharp internal angles in castings cause cracks.') They were designed to be as pleasing to the hand, as to the eye. 'That sort of bluntness of edge which long use confers on objects much handled he considered desirable to impart to the part of mechanism with which the hand or person came into contact,' Nasmyth told Samuel Smiles.[8] Like a fine prose stylist, Maudslay's skill was in knowing what to leave out, as much as what to put in. '[His machines] have an economy, a proportion, an elegance, coming from an art which, though perfectly functional, yet also contrives to be beautiful,' writes one engineering historian.[9] 'They were geometrically true and constrained to move in that geometry'. They did not rely on their heft for stability, and it was said that most of his machines were so neatly constructed that they would work upside down. There was a creative

---

7 Smiles, *Industrial Biography*, op. cit., 228.
8 Letter dated 29 January 1863, in 71075 BM.
9 F. T. Evans, 'The Maudslay Touch: Henry Maudslay, Product of the Past and Maker of the Future', *TNS*, vol. 66, 154.

exuberance to his work, a sheer diversity of inventiveness. He was a successful businessman, but unlike for some of his disciples, commerce was secondary to mechanical curiosity.

One early invention was a bench micrometer capable of taking measurements down to one thousandth of an inch. This machine was nicknamed the Lord Chancellor, as it was the Court of Final Appeal in his workshop. It was a sort of vice that held the object being measured between two perfectly flat steel surfaces. It worked by using a precision screw to bring the steel blocks closer and closer together until they touched both ends of the object being measured, at which point its size could be read off on the scale. When it was tested in 1959, the machine proved to have a mean error of an extraordinarily tiny 0.00079 inches. It dates from an age when most engineers measured to 1/16th or 1/32th of an inch. The idea of using a screw to measure length was not completely original – James Watt might have developed a machine along the same lines and micrometer screws were widely used in astronomy. But the level of accuracy attained by Maudslay was unprecedented in a practical industrial context, and paved the way for the development of modern precision engineering.

To mix metaphors, the screw is a building block of the modern industrial world, and one that is easy to take for granted. Stating the obvious, perhaps, the humble screw allows structures to be held together, and to be dismantled again. Screws could be tiny, or enormous, depending on the application. Some of the finest engineering minds of the early nineteenth century were preoccupied with the question of how and in what size to make screws. As Charles Holtzapffel, a London engineer, observed in Volume Two of his seminal but nowadays not frequently consulted account of *Turning and Mechanical Manipulation*, the science of screws was in 'a very imperfect state'. You had to make them yourself, using tools called taps and dies, which themselves had to be fashioned by hand, using hammer, chisel and file. So each screw was unique and could only fit into one corresponding nut. For the Portsmouth machines, for example, the nuts and bolts were numbered so it was possible to work out which fitted with what. In his unrelenting quest for the perfect screw, Maudslay fashioned taps and dies in multitudinous shapes and size, gear that attracted the wonder of contemporaries for its elegance and innovation.

In his youth, Maudslay had spent thousands of hours trying to build a *machine* to make the perfect screw. As early as 1800, he had developed a small, hand-powered lathe modified to cut screws. Decades later, he was still at it, building steam-powered machines based on the same principle. In 1829, the first task given to Nasmyth when he started at the works was to make modifications to what was called the origination machine. This created the *ur*-screw, the 'only begetter' as it were of an infinite succession of subsequent screws. The first step was to place a cylinder of hard wood or soft metal into a metal tube. There was a hole in the side of the tube, through which Maudslay could stick a crescent-shaped knife as the tube was rotated, thus creating the helix, the three-dimensional curve that wraps itself around the screw. Then a second cutter would be applied to the first incision and the thread would be deepened. Once the wooden screw was thus made, it could be transferred to the screw-cutting lathe, where it would serve as the master for another screw, identical in dimensions but this time made of steel. The crucial variables were the pitch – the interval between the threads – the diameter, the shape and depth of the thread in relation to the diameter of the screw, and whether it should turn to the right or left. As Nasmyth said, this machine became 'in the hands of its inventor the parent of a vast progeny of perfect screws, whose descendants, whether legitimate or not, are to be found in every workshop throughout the world, where first class machinery is constructed'. Soon after Nasmyth was put to work in the Maudslay factory, he made a scale model of a pair of 200 hp marine steam engines, which required about 300 minute nuts and bolts that had be reduced to the proportionate size. The screws had small collars, and Nasmyth built the machine to make these tiny screws.

Maudslay experimented with hundreds of different sizes and shapes of screws, creating small ones as well as the giant one, 5 feet long and 2 inches in diameter with 50 turns per inch, that is, 3,000 in all, exhibited at the Society of Arts. The nut belonging to this screw was 12 inches long and had 600 threads. Screws of similar enormous dimensions were at the heart of the Lord Chancellor and other micrometers used to make the most delicate measurements, for example as a dividing scale for astronomical instruments at the Greenwich Observatory. 'Divisions were produced [by its means] which could only be made visual by a microscope,' said Nasmyth.[10]

Precise observation was at the heart of the Enlightenment, a crucial adjunct to scientific endeavour; but like so many of the figures in this book, Maudslay was a working man who experimented his way to mechanical perfection. He

---

10 While Maudslay was developing and producing screws for the world of the engineer and manufacturer, others were deeply engaged in producing precision-made components for scientific purposes such as surveying, astronomy, horology and navigation. For further details see Randall C. Brooks, 'Towards the Perfect Screw Thread: The Making of Precision Screws in the 17th–19th Centuries.' *TNS*, vol. 64, no. 1, 101–19. Thanks to John Ditchfield for pointing this out.

was a *fabricant* rather than a *savant*, a maker rather than an intellectual: but the distinction between scientist and practical engineer is a modern one. The word 'scientist' was coined only in 1833 (by the philosopher William Whewell), after Maudslay's death. As with James Watt and other pioneering Lunar Men of the eighteenth century, there was a glorious blurring of intellectual curiosity directed at the natural world, and practical invention as dedicated to the needs of industry or war. Watt himself considered that he was carrying out chemistry experiments when he developed his steam engine, though of course he attained great mastery of practical engineering when it came to making his engines commercially viable. Josiah Wedgwood's understanding of chemistry led to great practical breakthroughs in the ceramics industry. Maudslay's empirical approach to technical innovation was typical of this phase of the Industrial Revolution. He and so many other inventor-entrepreneurs worked by trial and error. 'When one thing does not do, let us try another,' as James Watt succinctly put it. There were practical problems to be solved, for example the cutting of metal to appropriate levels of accuracy, and they were overcome not with a flash of startling originality, or as a result of new scientific understanding, but incrementally, by modifications of what already existed. He took what he knew worked for making small objects like watches, clocks and locks, and applied that, first to woodworking, and subsequently to working in metal.

This is not to imply that Maudslay was uninterested in science or natural philosophy: he was a man of great intelligence and intellectual curiosity whose friends included not merely Brunel and Bentham, but Michael Faraday, another man from a humble background, whose pioneering work in chemistry and physics led to the invention of the electric motor, generator and transformer. Together with Watt, both Maudslay and Faraday figure in a famous, mid-Victorian portfolio portrait of 'Distinguished Men of Science of Great Britain Living in 1807–8,' published by William Walker in 1862, and created over a period of years by several artists, including Sir John Gilbert of *Illustrated London News* fame (Figure 3). The picture, compiled from original portraits and photographs, shows an imaginary gathering of scientific and industrial luminaries grouped, as they never would have done in life, in the Library of the Royal Institution, at an arbitrary point in time. They include Sir Benjamin Thompson, aka Count Rumford, Samuel Bentham, William Murdoch, John Rennie, Marc Brunel, Charles Tennant (the chemical engineer) and Peter Dollond, pioneering optician. Sir Humphry Davy, professor of chemistry at the Royal Institution and pioneer of the miner's safety lamp, is represented, but his proletarian adversary (and arguably the more distinguished man), George Stephenson, is not (Davy and Stephenson had a public spat in 1816–17 over

FIGURE 3 The Distinguished Men of Science of Great Britain Living in the Years 1807–8, engraved by Walker and Echel, 1862 (Courtesy of Science Museum, London)

the safety lamp, which both invented independently, though great scepticism was cast on Stephenson's claim). For all its anomalies, the picture shows a mid-century appreciation of the melange of influences that gave rise to the Industrial Revolution, from pure science to practical mechanics.

Also in the engraving is Bryan Donkin (1768–1855), another supremely talented engineer who took out a number of patents and worked with his friend Maudslay on the problem of standardized measurements. Donkin was a pioneer of the fiendishly complicated printing machinery that was transforming the market for newspapers and popular literature, as well as the man who invented the fountain pen, and tin cans as a means of preserving food. Donkin took out a patent for canned food in 1812, and his tinned mutton, beef and carrots were sampled by the Prince Regent and the Queen, to their great satisfaction, and supplied to the Royal Navy. He had a factory in Bermondsey, in South London not far from Maudslay's, and Maudslay considered him a rare peer among engineers. Other frequent visitors to the factory were John Barton, the Master of the Mint (and inventor of the 'atometer,' a device supposedly able to measure the dimensions of an atom), and Frances Chantrey, the leading portrait sculptor of the Regency era, both of whom had a professional interest in the techniques of precision engineering. Towards the end of his life, Maudslay became fascinated by astronomy, commissioning Nasmyth to build a telescope.

For Maudslay, himself, practical advances in mechanical engineering did not depend on scientific understanding. A scientific explanation of the properties of the true planes developed by Whitworth in Maudslay's factory would come long after both men's deaths; a lack of understanding of the physics and chemistry of iron and steel did not stop Victorian entrepreneurs and their predecessors figuring out how to make the material stronger, more flexible and durable. A great leap forward had come in 1783–84, for example, with Henry Cort's discovery of puddling, a means of stirring molten pig iron to remove impurities, but the properties of wrought iron (as opposed to cast iron) were not properly understood, even as the material was used to construct increasingly sizeable buildings, bridges and ships. The maths and physics of tubular structures were not properly grasped, even as William Fairbairn was constructing gigantic bridges using huge iron tubes. Whitworth would revolutionise the accuracy of firearms by a similar process of trial and error. The properties of steam itself were not bottomed out until the 1840s, following the experiments of James Joule, the Manchester physicist and brewer. The lack of theoretical understanding did not stop practical men making steam the prime motive power for British industry long before Joule's work led to the formulation of the laws of thermodynamics.

* * *

An advertisement from 1812 demonstrates the range of products the factory was turning out in the years before the Battle of Waterloo:

> Henry Maudslay & Co, Engineers, London, beg leave respectfully to acquaint Gentlemen, Merchants, Manufacturers and their Agents, that they are enabled by their extensive Manufactory and Machinery at Lambeth, to furnish (upon reasonable terms) the most approved and complete Steam Engines, & when to send abroad provided with all necessary duplicates &c. of the wearing parts to ensure their perfect success in countries where mechanical assistance cannot easily be procured. They also beg leave to state that they keep on stock small engines and the principal parts of those larger powers most in demand that an order may be executed at the shortest notice. Millwork, Water Works & Machinery of every kind executed in their usual style of workmanship.[11]

---

11 Cited in Cyril Maudslay and Foster Petree, *Henry Maudslay, 1771–1831* (1948), 23.

The emphasis on replacement parts is new: few companies in the UK had attained the level of precision to produce an exact copy of a component. Likewise, producing for stock was very unusual and shows that Maudslay had figured out for himself the importance of economies of scale: manufacturing in bulk brought the cost of each item down and had the potential to make a bigger profit on each item sold. The risk lay in having to hold items as stock while waiting for a buyer to come along.

These calculations are still at the heart of modern process engineering and we see how Maudslay approached the complex interdependencies of cost, sales price and volume when, some time after 1810, the Naval Board approached him to manufacture iron tanks for carrying fresh water in ships. This was not his line of business, and he was at first nervous about taking on the work. But, keen to oblige an important client, he undertook to make one tank as a trial. Each tank needed 1680 rivet holes punched into it, a job that for the prototype had to be done by hand using presses. This was laborious work and meant that the first of the tanks cost seven shillings. This was a very large amount of money, which the Navy was unlikely to pay and on which Maudslay was not going to make a profit. However, the Navy proposed that he make 40 a week for an indefinite period. 'The magnitude of the order made it worth his while to commence manufacture, and to make tools for the express business.' He said to the Navy that if they ordered 2,000 tanks, he would supply them at the rate of 80 per week. He received the order and, through using specially constructed punching machines to make the rivet holes, he brought the cost of production down to ninepence per item. 'He supplies 98 tanks a week for six month, and the price charged for each was reduced from seventeen pounds to fifteen,' noted the admiring Charles Babbage.[12]

Despite this venture into mass production, Maudslay himself remained a generalist, putting his supremely inventive mind to work on a series of one-off challenges, always drawing on his keen understanding of precision engineering. He built the forming machine for making rope at the Chatham Dockyards – which machine was installed in 1810 and still functions to this day, with a claim to being the oldest working machine in its original location (it made the rope used on the set of the film *Pirates of the Caribbean*). Maudslay made sawmill machinery for the Woolwich Arsenal, and casting, rolling and drawing machines for the Royal Mint. This was a branch of manufacture that demonstrated enormous trust in Maudslay's ability, requiring perfect

---

12 Cited in Maudslay and Petree, op. cit., 81.

accuracy and precision as coins had to be uniform and of a quality that would be impossible to forge. He supplied similar machinery to the Royal Mint in Bombay, the Royal and Calcutta Mint, the Imperial Ottoman Mint in Constantinople and to Mexico – rare export orders at a time when selling machinery overseas was still strictly curtailed by wartime-era legislation. The factory produced calico printing machines for textile manufacturers, and later the curious Time Balls 'used to set ships' chronometers for the determination of longitude.' As Richard Maudslay (a descendant of the master) explains, these 'were installed in locations which were visible for ships in harbour. The ball was raised to the top of a tower and released usually at 1 p.m.' Bizarre-looking originals can still be viewed at the Greenwich Observatory or at the top of the Nelson Monument on Calton Hill, Edinburgh, and more than 100 were in place around the world.

Other inventions included the differential motion for raising weights (patented together with Bryan Donkin in 1806), the cup leather that sealed the joints of hydraulic presses, and a machine for regulating the supply of water to boilers at sea. He supplied his friend, Marc Brunel, with machines for his veneering factory, and also made bronze foundry equipment for Sir Francis Chantrey, the sculptor. There is little evidence that he sold machine tools, other than beautifully crafted, small lathes with slide rests intended for rich hobbyists, costing the substantial sum of £200 apiece. (Turning wood, ivory or metal had long been a hobby for the royal and upper classes, including King George III, of whom it was said he could have earned a reasonable living as a skilled craftsman). Later, Roberts, Nasmyth and Whitworth were celebrated for constructing the machines that built other machines, as customers came to their factories, saw the planing or slotting or boring machines at work, and placed orders for themselves. Demand for these capital goods took off in the mid-1830s following the coming of the railways and the near full mechanization of the textiles industry.

The small engines described in the advertisement were at the forefront of the new generation of engines developed in the United Kingdom around the turn of the century. For more than thirty years until 1800, the Birmingham firm of Boulton and Watt had ruthlessly enforced the patent, giving it monopoly rights over the production of their efficient but low-pressure engines, fighting legal battles with Richard Trevithick, Matthew Murray and others who sought to improve the original model. The threat of legal action eventually forced Trevithick in an entirely different direction, pioneering the use of so-called strong, or high-pressure, steam, which delivered far more power for a given size, but was considered exceedingly dangerous, as boilers and engines had a habit of blowing up under the high pressure. Such explosions remained a fact of industrial life for decades, and nineteenth-century newspapers abound in grisly descriptions of workers being roasted alive in factories and in ships. When Boulton and Watt's machines came off patent in 1800, the market opened up. While Maudslay and others were reluctant to adopt strong steam, they sought ways to build smaller engines that produced more power than traditional beam engines. In time, this allowed the development of steam locomotives, as it was impractical to have an old-style engine sitting on top of a carriage – though some railway pioneers tried it. For ships, the basic technology was modified so that the centre of gravity was lowered and the new engines had a side lever rather than a beam.

Maudslay developed a table engine, so-called because it sat on an elegant iron table, borrowing design features such as Doric columns from the block machines made for Portsmouth, and occupying very little space (Figure 4). As Ben Russell of the Science Museum has explained, the objective was to get rid of the rocking beam that had been a feature of steam engines since Newcomen's day, to reduce the weight of the machinery, and to lower the centre of gravity and make its workings self-acting – the cylinder sitting on the table, and the piston turning the crank and fly-wheel beneath. These engines were used the length and breadth of the country to drive machinery in mills and factories. Maudslay took out a patent in 1807, and the engine and its successors (made in a range of sizes, producing between 1.5 and 40 hp) became a staple product of the Maudslay factory for 50 years. Matching the stunning longevity of the block-making equipment, two of these engines powered the Maudslay works until 1900, and the one now housed at the Science Museum was in use from 1840 to 1948.

Maudslay was an engineer of genius, but his record is not unblemished. At 11:30 on the morning of 24 May 1826, the inhabitants of Lambeth were shocked by a tremor of the earth and a crashing noise 'as violent as that which would be produced by an earthquake'.[13] It transpired that part of his factory had collapsed under the weight of a newly erected iron roof. The use of iron for constructing buildings was in its infancy, and here there was a terrible miscalculation of the load-bearing capacity of the brick walls. It was thought that 100 workmen were at their posts in the engine shed when it collapsed, and crowds of anxious relatives assembled around the factory, waiting for news of their love-ones. This proved an exaggeration, but still women and children

---

13 *The Times*, 25 May 1826.

FIGURE 4 Maudslay's elegant table engine, 1815 (Courtesy of Science Museum, London)

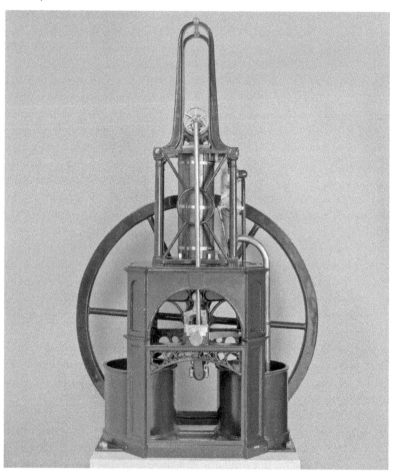

started to scream as bodies were brought out from the ruins. William Glover, domestic servant to Joshua Field, was standing near the 30-foot high wall adjacent to Francis St., cleaning knives, when it collapsed, and he was killed outright. Three more bodies were dug out of the rubble of brick and iron, and of the 15 or so who were seriously wounded, another three died later in the day. Others managed to survive by hiding under the iron machinery in the shed, waiting to be rescued. Maudslay himself had been on the roof earlier in the day, directing operations, and he did not attract personal criticism – *The*

Times stated categorically that it would be 'unreasonable to impute blame to any individual in a case of this kind.' Yet this was a rare and tragic error of engineering judgment.

* * *

With his naval connections, and located a few streets away from the Thames, it was not surprising that Maudslay should become involved with the shipping industry as it entered the age of iron and steam. Unusually, the decisive technological breakthrough came not in Britain, but in the United States, where in August 1807 Robert Fulton's *Clermont* became the first commercially successful steamboat following its maiden voyage along the Hudson River from New York to Albany. Just as with steam trains, contemporaries were deeply sceptical that steam could be used to propel boats, and then horrified and enthralled when they first saw one of these vessels afloat. 'A monster moving on the water, defying the winds and tides, and breathing fire and smoke,' is how one contemporary described the *Clermont*. The first steam tug was in fact of British design, the *Charlotte Dundas,* which steamed along the Forth and Clyde Canal to Glasgow in 1803, and the *Clermont* itself was powered by a Boulton and Watt engine. In 1812, Henry Bell ran the *Comet* on the Clyde between Glasgow and Helensburgh, the first commercial steamboat service in Europe. Within four years, steamboats were active on the river and estuarial waters of the Tay, Forth, Avon, Yare, Trent, Tyne, Orwell, Humber and Ouse, as well as the Thames, and in 1814 and 1815 two steamboats made it from the Clyde to the Thames, one westward through the Irish Sea and around Land's End, and the other eastwards.

In 1819, the American *Savannah* managed to cross the Atlantic with the help of steam, albeit it very slowly and too expensively to be viable. It was said, tartly, that her steam engine did not excessively detract from her sailing qualities. Yet steam-powered ships could prevail against wind and tide, and the technology was self-evidently superior to sail. The industry grew rapidly. The first steamboat packet to be seen on the River Thames was the *Margery*, which set off from Wapping Old Stairs for Gravesend on 23 January 1815, initiating a new era for tourism. By 1822, when Maudslay's business partner, Joshua Field, compiled a list of all 142 steamboats constructed in the United Kingdom over the previous decade, the vessels had become big, fast and powerful (14 were over 200 tons, with 10 having 60 to 100 hp engines) and were conducting coastal and short sea-services – from Glasgow to Liverpool, for example, Holyhead to Dublin, London to Hull, Liverpool to Bristol (via

Dublin) as well as cross-Channel services from Dover to Calais and Brighton to Dieppe.[14] By 1830, there were 57 steam packets on the Thames, catering largely for day-trippers heading east to Southend or west to Richmond or Kew, and also delivering the post. It was a fragmented industry, with dozens of companies around the United Kingdom making both engines and hulls, but Maudslay's firm achieved a series of firsts, building the engines for the 112-ton *Richmond* in 1815, one of the first steam-powered passenger vessels on the Thames, and for the *Regent* in the following year, and in 1821, and for the paddle steamer *Rising Star*, the first steamer to cross the Atlantic from East to West. These were private commissions, undertaken for entrepreneurs who could see the commercial potential of the new technology to open up longer and bolder passages. The Royal Navy, on the other hand, was highly sceptical, its commissioners declaring it 'their bounden duty, upon national and professional grounds, to discourage, to their utmost ability, the employment of steam vessels, as they consider that the introduction of steam was calculated to strike a fatal blow at the naval supremacy of the Empire'.[15] Yet by 1823, Maudslay supplied the engines for the *HMS Lightning*, the first steam vessel in the Royal Navy. In 1826, the *Enterprize* was the first steamship to make the passage to India, propelled by sail and by two Maudslay side-lever engines, with a combined output of 160 hp. The *Enterprize* took 113 days to complete the journey from London to Calcutta, far too slow to challenge the sailing ships of the East India Company, but fast enough to stimulate efforts to open up the longer sea passages to steam vessels. Two years later, Maudslay provided the engines for the Bombay-built *Hugh Lindsay*, the first steamship to travel in the opposite direction, from India to England.

Although the Birkenhead-based firm of Laird Brothers was pioneering iron-hulled boats from the 1820s, early steamship hulls were in the main made out of wood, the steam engine sitting inside, powering paddle wheels to the rear or the side. The practical problem was how to fit a heavy, vibrating and dangerous metal engine into the wooden structure of a boat. Maudslay addressed this by building some of the most powerful side-lever engines in use in the nineteenth century, and the firm also developed new technology such as the oscillating engine – compact 'self-acting' engines that were suited to smaller craft – and the so-called Siamese or double-cylinder engine. A key

---

14 John Armstrong and David Williams, 'The Beginnings of a New Technology: The Constructors of Early Steamboats 1812–1822', *TNS International Journal for the History of [Eng] and Tech*, vol. 81, no. 1 (January 2011): 1–21.
15 Denis Griffiths, *Brunel's Great Western* (1985), 8.

figure was his fourth and youngest son, Joseph (1801–1861), who was trained as a shipbuilder before he became an engineer and went on to take out 22 marine patents of his own. Together with Joshua Field, Joseph Maudslay built the firm into one of the leading marine engineering operations of the nineteenth century, their engines renowned for their power and reliability.

Maudslay's marine engines functioned as smoothly as the machines of the block factory, and to take a pleasure trip in one of the early steamships was evidently a delight. 'As we sped along we admired the ancient cedars, which gave dignity to the Bishop's grounds, on the one side, and the elms, laburnums, and lilacs, then in full bloom[; ...] the beautiful views that came into sight as we glided up the river, kept my father and brother in a constant state of excitement'. Thus, James Nasmyth described a cruise on the Thames aboard the Maudslay-built *Endeavour,* a small boat that plied between London and Richmond until 1840. But as the vessels became larger and more powerful, and their proprietors more adventurous, journeys were not always so congenial. Maudslay provided two 100 HP engines for the *Curacao,* the first steamship of the Dutch navy, which set sail from Rotterdam to Guiana in South America on 26 April 1827. There were mechanical problems galore. The boiler leaked, and saltwater quickly corroded the working parts of the engines. The floats on the paddles frequently broke. As the ship consumed coal, it rose in the water, which meant that the diameter of the paddles had to be increased. Her captain complained most vehemently of the vessel's incessant lurching. On 24 May, the ship arrived at the port of Paramaribo after what must have been an exceedingly uncomfortable passage of 4,000 sea miles completed in 28 days at an average speed of 6 knots.

By the time of his death in 1831, Joseph Maudslay's firm had supplied engines for 45 ships, and had built on speculation a pair of massive 200 HP engines, the biggest of their day, which were eventually installed in the HMS *Dee,* an 800-ton frigate commissioned by the Royal Navy. Maudslay was so proud of them that he commissioned the young Nasmyth to build a scale model, found to this day at the Science Museum. Crowds lined the streets of Lambeth when the immense components for the engines were transferred from the factory to the river for shipment. While most ships' hulls were still constructed of wood, the first ironclad steamer to be built on the Thames was said to be Maudslay's *Lord William Bentinck,* a 125-foot-long paddle steamer constructed for the East India Company for use on the Ganges. The vessel was launched in the summer of 1832 from the firm's dock at Pedlar's Acre, on the site of what became the Greater London Council offices. Engines from Maudslay's factory powered the *Rhadamanthus,* a hybrid schooner cum steamship built in wood at Devonport for the Royal Navy. Along with the *Dee*, the *Rhadamanthus* was one of the largest and most powerful ships of the age. With three masts and a standing bowsprit, and two 20-foot diameter paddle wheels, the 205-feet long, 813-ton ship was 'beautiful to look at, and as sharp at the bow as the bill of a snipe'. She carried four guns and 60 men and was the first Royal Navy steamship to cross the Atlantic, which she did in the spring of 1833. She steamed from Plymouth, across the Bay of Biscay, and then the Maudslay-built side-lever engines were disengaged and she sailed on to Madeira, where she refuelled before using a mixture of steam and sail to continue the journey to Barbados, arriving on 17 May.

The *Rhadamanthus*'s journey of 2,500 miles was covered at an average of just more than 6 knots, not much faster than she could have achieved under sail alone. But entrepreneurs and engineers could sense the enormous potential of taming the oceans. Already, many companies were offering scheduled sailing services across the Atlantic: at least, the departure times were scheduled, but when and whether or not you would arrive on the other side of the ocean, was entirely unpredictable. One in six of the earlier packet ships was shipwrecked, and those that made it could take as long as 54 days, compared to the standard time of just over 22 days for the Liverpool to New York run under sail. Amid considerable scepticism from the public and self-appointed experts such as the popular scientist Dionysius Lardner, who declared that ships would never be able to carry enough fuel, the race was on to build a steamship large and powerful enough to cross the Atlantic reliably and to a timetable that would attract paying passengers. Only five years later, the contest was won when Isambard Kingdom Brunel's immense *Great Western* opened up transatlantic travel to the general public, again powered by a combination of sail and Maudslay engines: as the French say, *à voile et à vapeur*.

CHAPTER 3

THE MAUDSLAY MEN

Beyond the advance towards mechanical perfection at sea and on land, Henry Maudslay's greatest legacy was his willingness to impart his knowledge to others. Like a Japanese *sensei*, Maudslay took pains to hire and train the best people to whom he could pass on the secrets of mechanical perfection. His protégés were talented men from humble backgrounds like his own, as well as those sent to the factory – as if to engineering finishing school. In the former category, two men stand out: Richard Roberts and Joseph Whitworth, two of the most inventive engineers of the nineteenth century; another, James Nasmyth would become a spectacularly successful businessman as well as a highly gifted engineer with a talent for self-promotion.

Roberts, later the creator of the Iron Man, the self-acting mule that revolutionized the textiles industry, was born in 1789 in the village of Carreghofa in Montgomeryshire, so precisely on the border between England and Wales that it was said the front door of his childhood home opened into Shropshire, and the back into Denbighshire. His father was a toll keeper, and a part-time cobbler. Roberts received little formal education, working until the age of twenty at first as a boatman on the Ellesmere Canal, and subsequently as a quarryman. In the spirit of the self-made men whose careers were later celebrated by Samuel Smiles, his virtuoso whittlings attracted the attention of the local vicar, who helped him with his lessons, and he became well known for fixing lathes and making spinning wheels. Fellow villagers were so impressed by his first spinning wheel that they clubbed together to buy the youth his first set of tools. Somehow, he got himself a job at Bradley Ironworks at Bilston, in the Black Country, operated at the time by the executors of the late John Wilkinson. There, he picked up enough knowledge of turning and pattern making (the process of making moulds in sand into which the molten iron is poured) to get him a job in Birmingham. Then he went to Liverpool, where there was no work, and on to Manchester, where he arrived at dusk, very weary and very miry, as he told friends decades later. He found a job as a wood turner, but did not stay long, as he was drawn for the militia in his home parish of Llanymynech, a common fate for young men at the height of the Napoleonic Wars, when manpower was scarce. Seeking to evade the warrant officers he believed were on his trail, he and a couple of mates tramped the 200 miles from Manchester to London, where he talked himself into a job as a turner and fitter, first at Holtzapffels in Covent Garden and then at Maudslay's, 'the great school from which some of the first mechanicians the world has ever seen have been sent forth,' as the *Mechanics' Magazine* put it. He spent two years at the factory in Lambeth and, although there are no records of what he worked on, he clearly learned a great deal about mechanical engineering. Smiles wrote that he acquired 'much valuable practical knowledge in the use of tools, cultivating his skill by contact with first class workmen, and benefitting by the spirit of active contrivance which pervaded the Maudslay shops. His manual dexterity greatly increased, and his inventive ingenuity fully stimulated, he determined on making his way back to Manchester, which, even more than London itself, at that time presented abundant openings for men of mechanical skill.'[1] He went back north in 1816, when the war with France was definitively won, and there was no danger that he would be called into the army.

Joseph Whitworth, who was to become the most celebrated mechanical engineer of the mid-century, took pains in later years to obscure his own humble and troubled origins. Newspaper reports conferred on him a respectable middle-class upbringing: he was supposedly the son of a successful schoolmaster who gave him all the benefits of a private education. But that was obfuscation. For hardship and emotional deprivation, Whitworth's early years stand comparison with those of the fictional David Copperfield. Whitworth as born in 1803 in Stockport, then a malodorous industrial town six miles to the southeast of Manchester. His father was a loom maker with intellectual aspirations. The family home was a back-to-back terraced cottage in a dingy court called Fletchers Yard, not exactly a slum but close to a bleach works, slaughterhouse, leatherworks and textile mills, and thus a noxious and dangerous place to grow up. In 1812, when Whitworth was just nine years old, the town was one of the first in the industrial north to be targeted by Luddites: looms were smashed, mills and mill-owners were attacked and there was rioting in the streets. Bizarrely, the Luddite wreckers in Stockport wore women's clothing. The government imposed capital punishment, and nine

---

1 Samuel Smiles, *Industrial Biography*, 266.

Lancastrians were taken to Newcastle to be executed. However traumatic to be brought up amid such insecurity, for his father was surely a potential target for the terrifying transvestite rioters, worse was to come. Whitworth's mother died in 1814 at the age of 34 and, shortly afterwards, he and his younger brother and their baby sister were abandoned. The sister was given up to an orphanage, where she died, and the boys were separated and fostered out with distant relatives, never to see each other again. Their father left home to pursue a religious vocation as a Congregationalist minister, entering a training college near Bradford, thus giving himself the formal education that his sons never received. Later, he established himself as a Nonconformist preacher and teacher in West Yorkshire, but Joseph did not attend his father's academy – contrary to conventional accounts of his boyhood – and never spoke to him again.

As a boy of 13 or 14 Joseph got a job in a cotton-spinning factory in Derbyshire: early biographies suggested that he worked in a mill belonging to an uncle, but this appears to be a fiction that Whitworth did nothing to correct. Three or four years later, he had acquired enough basic skills as a mechanic to set out on his own to find work in Manchester. According to his biographer, Norman Atkinson, who painstakingly demythologized Whitworth's early years, in the summer of 1820 at the age of 16 he took a job with Crighton & Co., a leading manufacturer of textile equipment. He stayed for 14 months before moving eventually to a Lever Street mill, whose owners took him on and paid him the full wage. 'It was the happiest day of my life to be recognized as a journeyman and to be paid as such,' he said in a rare comment on this phase of his career. He had done exceedingly well for a man without formal education, but he desired to advance further and headed to London, still the epicenter of engineering excellence, where he was to spend eight years. He left for the South on 21 December 1824, his 21st birthday. Rather than walking or taking the coach (which he could not have afforded), he travelled by barge, hitching lifts down the canal system through Cheshire to Nottingham. En route, he met and fell in love with the illiterate daughter of a bargemaster from Cheshire; they eloped and soon were married. By the time Whitworth came to wealth and fame, he and his wife were estranged, and he preferred to draw a veil over this romantic and impulsive step, at odds with his professional reputation. He joined Maudslay's in Westminster Bridge Road in May 1825, as an ordinary bench fitter and turner, a time when some 120 men worked there. Roberts and Joseph Clement had already left, and Nasmyth was yet to join, so their paths did not cross until later.

While working for Maudslay, Whitworth took on and perfected Maudslay's work in creating accurate planes. A perfectly flat surface is conceivable in theory, but impossible to execute in practice. It became an obsession for Whitworth to pursue this holy grail of pre-Victorian engineering, and he must have spent the proverbial ten thousand hours scrubbing and scraping metal plates. Even his hard-working contemporaries thought this was very peculiar, and his single-mindedness in this case is testament to his extraordinary doggedness in pursuit of mechanical perfection. The resulting true planes became the reference point, the literal benchmark, to secure accuracy in the manufacture of machines such as steam-engine valves, printing-press tables, slides of all kinds and, above all machine tools such as lathes and planing machines used to construct other machines. 'All excellence in workmanship depends on having a true plane,' Whitworth would proclaim later in his career.

At first, he thought that to obtain a perfect flat surface, he needed two planes, each of which had to be worked and reworked until it fitted its counterpart with complete accuracy. 'If these planes were true, one of them ought to lift the other,' he reasoned. Whitworth was sitting at his bench next to a Yorkshireman named John Hampson. 'Thou knows nowt about it,' said the Yorkshireman. He covered the surfaces with ink, which would reveal undulations and protrusions and other blemishes undetectable to the naked eye, and these points could then be eliminated altogether with hard steel scrapers, until at last there was what appeared to be a perfectly flat surface. Then it dawned on him that two planes could stick together and still be imperfect, for example, if one were concave, and the other convex. In order to get to true perfection, three interchangeable planes are required. He took the planes home with him, scraping away at the surface of the plates. First numbers 2 and 3 would be matched to number 1, then numbers 2 and 3 to each other, and then finally number 1 to numbers 3, and so repeatedly, until absolute truth was attained. 'Ay, tha's done it,' said the Yorkshireman, on being presented with the result of the young man's labours. In his maturity, Whitworth used the same principle to develop right angles and cubes with true planes, using three blocks and making sure the sides of one cube were properly aligned with the sides of all the others. This was painstaking, tedious work, but it was the basis for creating perfectly accurate standard measurements and gauges.

James Nasmyth was born into an entirely different milieu from that of Maudslay, Roberts or Whitworth. Indeed, while he admired Roberts, his relations with Whitworth were famously strained in later life, perhaps because of the difference in social standing as much as professional rivalry and different conceptions of the purpose of mechanization: Whitworth was an idealist, believing that advances in technology would bring greater prosperity to the masses, while Nasmyth was cynical and commercial, seeing the new machinery as a means of reducing labour costs and dependency on unreliable working men. His father was Alexander Nasmyth (1758–1840), a celebrated Edinburgh artist, best known today for his portrait of Robbie Burns in front of the Brig o'Doon in Ayrshire, and for the landscape paintings that earned him the sobriquet the Claude of Scotland. His son was named after James Hall, his friend and president of the Edinburgh Royal Society: testament to powerful social and professional connections in the Athens of the North. Sir Walter Scott and Henry Raeburn, the artist, were close family friends, and much of the Scottish nobility were the artist's clientele. Indeed, Nasmyth's mother was the daughter of a baronet, so the future engineer had a family connection with the lower reaches of the aristocracy. James thus grew up on familiar terms with writers, painters, politicians and peers, all frequent visitors to his parents' house in York Place in Edinburgh's elegant New Town. More importantly for James's future career, his father was a hobby engineer and inventor, with a forge in the family home. (His standing as a man of science was considerable: he was later included in the portrait of Distinguished Men of Science of Great Britain Living in 1807–8, along with James Watt, Maudslay, Donkin, Trevithick, Sir Joseph Banks and others. See p. 31.) The family's circle of influential acquaintances included Watt, who paid a visit in 1817, when nearly 82 years of age. 'It was but a glimpse I had of [Watt],' recalled Nasmyth. 'But his benevolent countenance and his tall but bent figure made an impression on my mind that I can never forget. It was […] something to have seen for a few seconds so truly great and noble a man'. For all these connections, the Nasmyths were always short of cash, and there is much of the self-help paradigm in James's life story.

All told, there were 11 siblings, the eldest of whom was Patrick, a talented and unworldly artist best known for the manner of his death in August 1831 at the age of 44, rather than for his paintings. Having spent the day sketching some picturesque pollard willows on the bank of the Thames, his feet planted in waterlogged ground, Patrick caught a severe cold and was confined to his lodgings in South Lambeth, close to the Maudslay factory. A few days later, there was a tremendous thunderstorm. As the rumbling stopped, the clouds dispersed and the setting sun burst forth in a golden glow. The artist asked to be propped up in bed so he could memorise the dramatic effects. His wish was granted, and then he died. Another brother, George, shared many of Nasmyth's early entrepreneurial adventures and later became his business partner. There were also six talented sisters, all gifted painters. They assisted their father in arranging drawing classes, held in the artist's large studio at the top of the house, helping shore up the family's finances, and they exhibited widely. Four of the sisters did not marry and were based chiefly in London, where they lived long, independent and successful lives. 'They all continued the practice of oil painting until an advanced age,' Nasmyth wrote. James himself was a talented amateur artist and his skill as a draughtsman certainly informed his engineering. The Institution of Mechanical Engineers in London holds a collection of his finely executed line drawings for the steam hammer, as well as a portfolio of more whimsical studies, for example, of fairies.

Unlike George Stephenson, or Henry Maudslay himself, Nasmyth had the opportunity to go to school, but he did not take to book learning and was beaten up by one teacher for being unable to learn his spellings, his head banged against the wall of the classroom. He spent three years at Edinburgh High School but did not take to the classical education on offer there either. A highly intelligent boy, he could not cope with formal learning and was happiest working with his hands. He spent his free time in his father's workshop, making spinning tops for his school friends. A friend's father owned an iron foundry in Edinburgh, and he spent his Saturday afternoons there, lending a hand with the repairs of millwork and steam engines. 'I did not read about such things,' he wrote later, 'but I saw and handled, and thus all the ideas in connection with them became permanently rooted in my mind.' Nasmyth would always maintain that "words were of little use" in this form of practical education. Eyes and fingers were the only true inlets to sound practical instruction, he believed. Like many other engineers of the period, including Brunel himself, he was extremely sceptical of the value of theoretical learning and had no faith in engineers who wore gloves.

Nasmyth turned his bedroom into a forge, where using his fireplace he taught himself to cast iron, weld and practice smithery, the heat and explosions causing much consternation in his refined household. By the time he was a teenager, he was constructing little brass cannon and working models of steam engines, which he sold to the university for £10 apiece. He paid a third of the proceeds to his father by way of rent and used the rest to purchase admission to the Edinburgh School of Arts for lectures on chemistry, mathematics and geometry. Through his father's formidable connections, he was befriended by Robert Leslie, professor of natural science, and Robert Bald, a celebrated mining engineer. In 1827, when he was 19 years old, Nasmyth and his brother, George, made a small model of a steam carriage to run on a common roadway. This attracted the attention of the Scottish Society of Arts, which subscribed £60 for him to build a full-scale carriage capable of carrying four to six people.

The resulting contraption took him four months to build and made several successful journeys along the Queensferry Road.

Having made his way to London, it was a matter of perpetual pride for Nasmyth that he was hired as Maudslay's personal assistant, rather than as an apprentice, and that he was paid a pittance at least. In his first year, he was paid 10s a week, and in the following year, this was raised to 15s, and only then, he says, 'I began to take butter to my bread'. He cooked his own dinner in a stove of his own invention. Decades later, when rich and retired, he could not help bragging to a group of MPs how he had saved money by walking to Farringdon Market to the north of the river, where he could buy meat for 2.5d a pound compared to 3.5d at home in Lambeth. The heat for his slow cooker came from a single wick lamp that he would light on leaving for work in the morning. The meat 'was stewed to perfection in my apparatus', he told the parliamentarians. The device signalled his spirit of self-denial and his love of true independence, and was the real foundation of his fortune, he declared.

An older man who spent time at the Maudslay factory was Joseph Clement (1779–1844), the son of a Westmorland handloom weaver. Clement had come to come to London in 1813 at age 34, with £100 in his pocket and a wealth of practical experience in the mills and factories of the North. He got a job as a mechanic and then progressed to become works manager at Joseph Bramah's (by this time in Pimlico) and then to chief draughtsman at Maudslay's in Lambeth. A mere six years younger than Henry Maudslay, Clement was supremely skilled in preparing technical drawings and was put in charge of designing the firm's marine engines. Confident of his own abilities, he left Maudslay's in 1817 to set up his own workshop in Prospect Place, at Newington Butts near Elephant and Castle in Southwark. Here, he made a fortune as manufacturer of the components for Charles Babbage's Difference Engine.

Other Maudslay men were the sons of the early industrial aristocracy, including most obviously Isambard Kingdom Brunel, who was sent there part-time by his father as part of his tailored engineering education. Another was David Napier, a scion of a great Scottish shipbuilding family: he arrived in London from Glasgow in 1810 and, after working at Maudslay's, set up his own works in Fetter Lane, specializing in elaborate printing machinery. Later, he tried to build a road-going steam carriage, with little success; he is easily confused with two other cousins called David Napier, one who opened a factory in Lambeth, and another who became a celebrated but controversial shipbuilder, as his marine engines had a habit of blowing up. Samuel Seaward took an apprenticeship with Maudslay after two years at sea as a midshipman;

he then worked with the great Richard Trevithick at the Hayle Foundry in Cornwall before setting up in business with his brother, John, at the Canal Ironworks in Millwall, East London. Their company became highly regarded as general and marine engineers, inventing the direct-acting paddle engine known as the *Gorgon*. William Muir, a talented Scotsman was a nephew of William Murdoch, James Watt's right-hand man. He worked for Maudslay before joining Whitworth in Manchester. Muir fell out with Whitworth after the latter expected him to work on Sundays. In turn, he established his own business and is credited with making the first railway ticket machine, based on the pioneering design of Thomas Edmondson.

Like the heroine of Jane Austen's *Emma* (1815), these young men would have amused themselves by a visit to Astley's around the corner, or Vauxhall Gardens in nearby Kennington. Joseph Whitworth kept dogs, while Muir, a stern Scottish sabbatarian, spent much of his leisure time attending chapel. For all the distractions of Regency London, they were, in the main, a hard-working, earnest group, bent on self-improvement. William Fairbairn, later one of the great engineers of the period, first arrived in London around 1810, eventually finding a place at John Penn's ship works in Greenwich. As a working man, his experience would have been similar to that of many at Maudslay's: he went to the theatre once or twice a week, attended public lectures and read as much as he could: some Smollett and Fielding for amusement, but for edification, he absorbed the first three chapters of Euclid's *Elements*. Decades later, when he entered into a very public argument over the construction of the tubular bridge over the River Conway, he demonstrated a knowledge of mathematics and geometry that rivalled, if not exceeded, that of the best university professors of the day. The Maudslay men would have read as much as they could about the science and mathematics underpinning the practical work they were doing by day. Some attended the early meetings of the Institution of Civil Engineers, established in 1818 by Joshua Field and Joseph Maudslay, Henry's youngest son, and others including the gifted Bryan Donkin.

At first, this was a group of enthusiastic young men who met in the evening to discuss mechanical subjects. They would exchange 'useful information [...] spread for the general good [...] free from restraint of commercial interests, or rival jealousies, vying only in their love of the subjects brought before the meetings'. A few years later, the engineers brought on board as the first president the celebrated Thomas Telford, the so-called Colossus of Roads, the builder of harbours, bridges and canals, thereby conferring dignity on a group that by then numbered no more than sixty. In 1768, the term *civil engineer* had first been used by John Smeaton, one of the eighteenth-century giants, meaning someone who was not employed by the military, as opposed to the modern meaning of an engineer who designs and builds roads, tunnels, bridges and the like. In the nineteenth century, then, the first civil engineers were mechanicals. The mechanical engineers created their own institution only in 1847, based in Birmingham, with George Stephenson as their first president. Stephenson had refused, it was rumoured, to submit the necessary essay on engineering required to join the civils. After all, he was the great pioneer of railways, the most extraordinary engineering innovation in a thousand years of British history. Fed up with their increasingly gentrified peers in the South, the mechanical engineers of the Midlands and the North set up their own, proudly provincial, body. The two institutions now have more than 200,000 members between them and are located next door to one another in grand premises in Westminster, barely fifteen minutes walk across the River Thames from the site of Maudslay's factory.

These years after the end of the Napoleonic Wars were a time of ceaseless strife between workingmen and their masters, and of violent contention as the working classes fought for political representation in Parliament. As E. P. Thompson has written, this was the 'heroic age of popular radicalism'. Parliament was unrepresentative of the rapidly changing industrial society that Britain had become. Peace brought economic downturn as hundreds of thousands of troops were demobilised; wages fell and prices rose. The 1815 Corn Law protected the interests of landowners by imposing a tariff on imports of corn, thus keeping food prices high. Economic development was far from smooth: there were intervals of deep depression and economic crisis, for example: 1825–26, 1836–37, 1839–42, 1846–48. Early in the century, riots, strikes, arson and machine-smashing were commonplace, such as those no doubt witnessed by young Whitworth in Stockport. Now, amid poverty, unemployment and social unrest in the countryside as well as in industrial areas, there were fears the country would suffer revolution. Riots erupted in London in 1816 and an ineffectual march on the city by Lancashire weavers the following year. The most egregious example of class conflict was the Peterloo Massacre in the heart of Manchester on 16 August 1819 (see chapter 5). In 1820, the Cato Street conspiracy to assassinate the entire cabinet was exposed. Revolution was averted, unlike in France in 1830, but the social unrest continued (e.g., the agricultural Swing Riots of 1830–31) and the Great Reform Act of 1832 merely opened up the franchise to men of property and the cities of the North, leaving the Chartists to campaign in vain for mass suffrage.

But for all the suffering of the working man, for many it was a time of opportunity. There was a hunger for learning, catered to by popular lectures by scientists such as Humphry Davy and Michael Faraday in London and John

Dalton in Manchester, and poets like Samuel Taylor Coleridge, who filled the hall of the Royal Institution in London's West End. In 1823, George Birkbeck set up in London the Mechanics' Institute, the first of its kind, and a provider of evening classes to ambitious working men. In 1825 the *Mechanics' Magazine* was founded, a publication that contained detailed accounts of mathematical problems and engineering case-studies, as well as more eclectic subject matter appealing to the intellectually curious. Topics covered in one of the early editions included the poisonous nature of African trees; the general nature of wheels and springs for carriages; the idea of an air engine; mechanical geometry; the properties of cast iron colonnades; and tapping nuts and cogwheels. William Blake denounced the first issue of the magazine as 'an enemy to art', but fellow artisans were more enthusiastic: the magazine built a circulation of 12,000, its readers constituting the aristocracy of manual workers. There were mechanics' institutes, adult-education colleges, publications and societies such one for the Diffusion of Useful Knowledge. In Henry Brougham's influential words, this was the time of the 'March of Progress', and the mass production of books, pamphlets, periodicals and newspapers was the means by which knowledge was disseminated and consumed. Historians and social scientists, from Karl Marx onwards, have tended to emphasise the dehumanising impact of increasing mechanisation, but the Maudslay men remind us of another side of the story, of the gratified quest for knowledge and a hunger for self-improvement. Their stories suggest that the knowledge and know-how, acquired on the job and through networks such as the engineering institutions, provided the intellectual capital on which enterprising engineers could build successful lives and businesses.

Fifty years after Whitworth worked at the factory – by now a baronet and one of the richest men in England – he confided in his old friend the politician John Bright, musing on the reasons for his success. Whitworth attributed it to his industry as a young man and 'anxiety to study; his resolve to instruct himself in everything connected with mechanics and machine-making.' Somewhat ruefully, perhaps, he told Bright that he had led a blameless life as a young man: 'I was a very good lad, never a better.' He had worked hard, studied, and conducted himself well. For many months he had lived on about five shillings per week for food. 'He is now very rich,' Bright noted after staying at Whitworth's mansion in Derbyshire. '[He] has no children, wishes to leave or give much of his property for public uses'.[2] Beneficiaries of his

---

2 See John Bright's diary for 12 November 1877.

scholarships had to demonstrate practical mastery of one or more tools such as the axe, the saw, the plane, the hammer, the file or the forge. They needed to show skill in handicrafts such as smiths' work, turning, filing and fitting, pattern making and moulding. Their first exam concentrated on theoretical subjects: mathematics (elementary and higher); mechanics (theoretical and applied); practical plane and descriptive geometry and mechanical and freehand drawing as well as physics and chemistry, including metallurgy. This was the hugely demanding combination of practical and academic skills he had mastered as a young man.

* * *

Henry Maudslay died at home in Lambeth on 15 February 1831, some months short of his 60th birthday, having contracted a severe chill after a dash across the Channel to France to visit a sick friend. He was buried in the churchyard of St Mary's Woolwich, in a cast-iron tomb of his own design. The epitaph on the side highlighted the precision and beauty of his work, and his personal qualities as a friend and mentor:

TO THE MEMORY OF

HENRY MAUDSLAY

BORN IN THIS PARISH 1771

DIED AT LAMBETH FEB 15TH 1831

A ZEALOUS PROMOTER

OF THE ARTS & SCIENCES

EMINENTLY

DISTINGUISHED AS AN

ENGINEER

FOR MATHEMATICAL ACCURACY AND

BEAUTY OF CONSTRUCTION

AS A MAN FOR

INDUSTRY & PERSEVERANCE

AND AS A FRIEND FOR A KIND &

BENEVOLENT HEART

'How his workmen loved him,' Nasmyth wrote. 'How his friends lamented him.' Samuel Smiles memorialized him as: 'The very beau ideal of an honest, upright, straight-forward, hard-working, intelligent Englishman.' Maudslay's legacy was immense and will be explored in subsequent chapters, starting with an account of the construction of the first tunnel under the Thames, considered one of the wonders of the age.

CHAPTER 4

THE THAMES TUNNEL –
A WONDERFUL UNDERTAKING

In the summer of 1816, Marc Isambard Brunel was contemplating a road tunnel underneath the Thames. The intention was to link the great centres of traffic and commerce to the south and east of the metropolis with the North. In those days, all goods unloaded from ships moored in the Pool of London, or at the London Docks and West India Docks on the north shore, and destined for the South, had either to be transported by water or be carted several miles through narrow and crowded streets and over London Bridge. Likewise, ships servicing the coasting trade, bringing coal and passengers down from the north-east, needed to decant their cargoes destined for the north of the river. There was no question of building another bridge to the east of London Bridge, as this would impede shipping, and Tower Bridge itself would not be built until 1894. As the Duke of Wellington put it, addressing one of many meetings of shareholders called to shore up the tunnel's finances: '[A]ll felt the benefits not only to its immediate neighbourhood and the populous districts around it, but to the counties generally of Essex, Kent and Surrey, between which it would open up a short and convenient form of communication'.[1] Initiated with the backing of the victor of Waterloo in the years immediately after the nation's triumph over France, the project held international political and military significance as well as spectacular commercial potential. Or, at least, so thought the duke and many others decades before the tunnel became a byword for delay and profligacy.

The undertaking was at the time a prospect more daunting than constructing the tunnel under the English Channel in the twentieth century. There had been earlier attempts to link the Thames close to this location,

---

1 *The Times*, 7 July 1828.

but the technological challenges had proved insurmountable. The gravel underlying the river was unstable, and it was impossible to dig deep enough avoid the risk of irruption from the river. The plan was to construct two carriageways of 1250 feet in length to be housed within a brick frame, but how could one advance through the gravel and erect the brickwork at the same time? This was the conundrum exercising Marc Brunel's supremely inventive mind in the summer of 1816. In one of those Eureka moments beloved by the biographers of the heroic engineers, Brunel happened to be staring at a block of wood in the Chatham Dockyard when his attention was caught by the actions of a *teredo navalis* shipworm that was boring its way quite successfully through the wood. The worm attached itself with its nose and drilled away using a pair of strong shelly valves, disposing of the detritus through its backside. Brunel at once determined to copy this natural design to create a tunnel-boring device. He took out his patent 'for forming drifts and tunnels underground' in 1818, and thereafter set about raising the enormous sums required to finance this ambitious project. By 1824, the public had invested the gigantic sum of £179,900, and the digging could get under way.

As in the case of the block factory or the Difference Engine, Brunel could conceive of the design, but he could not build the machine. There was only one place to turn to get the tunnelling shield built: Henry Maudslay's factory in Lambeth. The result of Brunel's commission was a contraption 22 feet in height, 37 feet wide and 7 feet deep: a gigantic, 140-ton iron shield, which began its work in late November 1825 after the access shaft at Rotherhithe had been completed. The device was divided into 12 perpendicular frames, each subdivided into three cells or boxes. The revolutionary principle was to conduct the actual digging through 36 small, rectangular windows, each one containing one or more miners. It inched its way through the mud and slurry under the Thames, creaking and cracking under the immense pressure of silt and water. The vertical and lateral pressure on the shield was calculated to be 3,000 tons. The edifice supported the weight of the earth above while the miners did their work, excavating the soil in front of their compartment and then putting in place small planks to secure the additional space. At the appropriate moment, the shield could be manoeuvred nine inches forward using a pair of horizontal levers, one on the bottom and the other on the top of the contraption. Once propelled forward, the brick workers would move in to build the wall of the tunnel, and thereafter the painstaking exercise could begin again. Behind the shield came bricklayers who built the double-arches of the tunnel thoroughfares, using quick-setting Roman cement.

In all, 500 men typically worked the tunnel at any one time, their shifts 16 hours one day and 8 the next, comprising not merely miners and bricklayers but blacksmiths, carpenters, riggers, millwrights and labourers. Everything needed for the construction was manufactured onsite. 'The work is carried on with great strictness as to time,' it was said. 'Fines are imposed for absence without sufficient excuse or drunkenness.' They would get through 70,000 bricks a week, 350 casks of cement, 300 lbs of candles and would cart 750 tons of spoil from the works. The company provided its workers with oilskin caps, thick flannel clothing and stout woollen stockings; those toiling in the bottom cells were given galoshes that covered the knees. An insurance fund was operated by deducting 6d a week from each man's wages, thus providing cover in the event of accident or death.

The new technology proved effective, and by late 1825, the 18-year-old Isambard Kingdom Brunel, given day-to-day charge of the operations by his father, hosted a dinner in the bowels of what then existed of the tunnel for fifty backers, celebrating the progress of the undertaking. All went well until the spring of 1827, as the tunnel approached the midway point of the crossing. The depth of the tunnel was an average of 80 feet below the surface of the riverbed, but as it made its way further to the south, the distance between it and the bottom of the river at its deepest point narrowed to a matter of a few yards. When the workmen opened up the doors of their cells to dig out the earth in front of them, it became obvious that they were not excavating solid earth, but a permeable substance with the consistency of slurry, disconcertingly close to the river itself. One engineer described the river mud as a 'gelatinous, offensive, compound of putrefaction, rotting and destroying wherever it rests'.[2] Sailors on the river above would amuse themselves by casting beef-bones, oyster shells and empty bottles into the Thames, and in due course these objects would turn up in the soil being dug out by the tunnellers. Brunel and his father were in the habit of descending to the bottom of the river in a diving bell, a brave and possibly foolhardy course of action. (The pressure of the water trapped the air inside the bell). On one occasion, they dropped a spade from the bell, and it turned up a few days later, in a mangled state, in the detritus in front of the tunnel shield.

Young Brunel had enjoyed, if not a pampered upbringing, at least the perfect education for a future engineer. Before he took on the tunnel, his

---

2 Cited in Michael Chrimes et al., *The Triumphant Bore: A Celebration of the Thames Tunnel* (1993), 11.

practical experience was limited to an internship with Maudslay (where he acquired 'all [his] early knowledge of mechanics') and a spell with Abraham-Louis Breguet in Paris, where he learned the delicate art of watchmaking. His sojourn in the tunnel seemed designed by his father to ensure that Isambard got not merely his hands dirty, but the rest of his body too. Throughout the spring of 1827, Brunel was constantly falling or jumping into the filthy water of the Thames, a semi-liquid sludge of mud and sewage. At about 6 p.m. on 18 May, for example, when 120 workers were engaged below, a 'portion of the earth gave way and the water rushed down in a torrent' from behind the shield. The workers fled towards the shaft in a state of understandable terror, bounding up the ladders five at a time. All but one made it out, and this fortunate soul was saved by Brunel, who jumped into the floodwater just as the man's strength was giving out. The next day, Brunel and his father were back in the diving bell, ascertaining that the cause of the leak was a mere perpendicular fissure in the earth that could easily be stopped by throwing bags of clay into the river and directing them to the best spot using long spikes.

A few days later, Marc Brunel was investigating the brickwork from a boat when one of his workmen moved too sharply and he was once again thrown into the water. This time, he and everyone else emerged safely, but this was not the case on Tuesday, 26 June, when a party of six businessmen and workers descended into the still-waterlogged tunnel to ascertain the level of damage. One man, Robert Humphrey Martin, stood up suddenly in the flat-bottomed boat in the hope of examining the brickwork in the vicinity of the shield. Described in contemporary reports as a 'very heavy man', Martin overturned the boat. He himself was close enough to the central pier of the tunnel to swim, or perhaps wallow, to safety. 'A sense of alarm and confusion now ensued that beggars all description', according to the report. It was *sauve qui peut* in the foul waters, still littered with the detritus caused by the flood, as those who had fallen from the boat struggled to find their way to niches in the brickwork. After a few minutes of panic, there was a roll call and it became apparent that one Samuel Richardson was missing. Once again, Brunel stripped off, diving repeatedly in and out of the water in the vain hope of finding the man. Richardson's body was recovered half an hour later. His widow came down to the tunnel in a distressed state, and could barely be persuaded to return home. Martin survived, though in a state of severe shock due to the impact of the water on his large frame.

For nearly five months, there was calm, as 30–40,000 cubic feet of water and mud were pumped out of the tunnel, and the hole in the bed of the river plugged with Portland cement and clay. The shield inched forward under the Thames.

'British will power coupled with Brunel's own genius and determination were such that no one was deterred by this misfortune,' wrote Johann Fischer, a visiting Swiss industrialist. 'All the resources of the human spirit, all the knowledge of science, and all the aid that money could give were united to triumph over the effects of this disaster, even at the risk of life and limb.'[3] This assessment proved optimistic. Shortly after 6 a.m. on Saturday, 12 January 1828, Brunel the younger was supervising the changeover of the excavators' shift just short of 600 feet along the tunnel, when he noticed that water was draining through the soil at the face of the shield much more rapidly than usual. He put it down to the pressure of the high tide above. The top of the shield was a mere 17 feet from the bottom of the river at this point. Half an hour later, as Brunel was standing with three other men in Box Number One at the top left, or western end, of the shield, the earth buckled in front of them. There was a sudden irruption of water through a hole no bigger than six inches across. Brunel had the presence of mind to step round into the neighbouring compartment and give directions to the other men to leave the shield. The men began to clamber down the wooden frame erected behind the shield itself. Brunel and three others had made their way twenty or so feet along the tunnel towards safety, the water already up to their waists, when the tunnel was plunged into darkness. Amid the terrifying din of men screaming and water bursting through from the river above, Brunel was shouting directions to safety when he and the other men were overtaken by a surge of water carrying the timber frame, forced from the back of the shield by the pressure of the current. Brunel was hit on his leg by the timber and dragged under water. Thomas Ball and John Collins, who had been with him in the box, were swept away and never surfaced. Jeptha Cooke, working in one of the bottom compartments, got behind the rush of water and perished, and three other so-called bottom box men were also killed.

On surfacing, the half-drowned Brunel's first instinct was to swim back to the shaft, and then he tried to turn back against the current in the hopes of saving the men. But the force of the water was irresistible and drove him, along with scores of others, to the bottom of the shaft, where they were dashed repeatedly against the end of the tunnel, unable to haul themselves out as the bottom end of the ladders had been smashed. Then the water forced itself up to the very top of the shaft, spewing forth men and detritus. Young Brunel 'had no alternative but to abandon himself to the tremendous wave, which, in a few seconds bore him on its seething and angry surface to the top of the shaft'.[4] Six

3 W. O. Henderson, ed., *J.C. Fischer and His Diary of Industrial England*, 30.
4 Beamish, op. cit., 262.

men died that morning, while Isambard Brunel himself was spat out of the foul torrent, semi-conscious and seriously injured. Crowds of desperate relatives, and the merely curious, gathered at the tunnel entrance.

The day after the accident, Marc Brunel descended to the bottom of the river in the diving bell, ascertaining that the rupture had been caused by a modest seven-foot-long oblong fissure in the bottom of the river. He received hundreds of letters from members of the public offering suggestions as to how to fill up the hole.

Meanwhile, Isambard was packed off to Bristol to begin a long period of recuperation. From his sickbed in the elegant suburb of Clifton, he looked out from the cliffs over the Avon Gorge and his hyperactive mind set to work. The city elders had launched a competition to build a bridge across the gorge. After proposals from Thomas Telford, the doyen of British civil engineering, were rejected, Isambard Kingdom Brunel won the follow-up competition with his design for what became the world-famous Clifton Suspension Bridge. This project, of course, was not completed until after his death. Brunel's Bristol connections led to his appointment in March 1833 as engineer to the Great Western Railway. He was not quite 27 years old when he took on the project that he intended to be 'the finest work in England,' He did not play any further role in the construction of the tunnel, which was in any case closed down in August 1828, after the project ran out of funds.

'Of my own knowledge I can speak of the interest excited in foreign nations for the welfare and success of this undertaking,' pleaded the Duke of Wellington in the summer of 1828. '[T]hey look upon it as the greatest work of art ever contemplated.' Neither the Duke's support, not Marc Brunel's eternal optimism, was enough to resurrect the tunnel project immediately. As with infrastructure projects of our own day, there had already been spectacular cost overruns, and investors' initial capital was exhausted. The tunnel was supposed to cost £200,000, and to take seven years to complete. Already, expenditure was close to this level, and the tunnel was barely halfway built and, in fact, it would take nearly twenty years to construct. At least another £200,000 was needed. Amid the squabbling of private shareholders, the necessary funds could not be found, and the government was unwilling to step in, with the result that for seven years, the Great Bore, as the tunnel was satirically named, was abandoned: a national disgrace for some, for others an appropriate fate for a project that could not pay its own way. Work started again in December 1834, only after the government advanced a loan of £247,000, an enormous sum that would never be recovered. (There was a new shield, this time constructed by Rennies.) This intervention from the state was a rebuke to the ruling ideology of laissez-faire. Unlike the investment in the block factory, there were no vital interests of national security, but politicians of all parties were persuaded that the prestige of the country was at stake: it simply had to be finished.

And finished, finally, it was: after years in which the tunnel advanced by 'just the breadth of one halfpenny per day,' as Brunel put it, subject to the usual flooding, collapses and irruptions of noxious gases, the tunnel reached the Wapping shaft in December 1841. On Saturday, 25 March 1843 it was formally opened – not to the general public or to commercial traffic, but to a crowd of entrepreneurial, scientific and social grandees, including Marc Brunel, Charles Babbage, Michael Faraday and various noblemen and their wives. They formed a parade, a brass band at its head, which led off from Rotherhithe to Wapping, where they climbed out for fresh air, before returning the way they came. The journey down the shaft was claustrophobic, the staircases rickety, the ground in the tunnel damp, the walls clammy and the flaring of the gaslights alarming, contributing to an atmosphere 'at once disgustingly heated and fetid'. Despite the reverberation of the music and the good-spirits of the holiday crowd, 'it was evident that there was a chilling lurking fear in the hearts of many' – and perhaps a sense of bathos? 'Well, it is a wonderful undertaking,' wheezed one citizen as he stood before the entrance.[5]

* * *

Queen Victoria was subject to eight assassination attempts, some of them serious. But there is one other occasion when the Victorian Age could have come to an abrupt and tragic conclusion. After lunch on Wednesday, 26 July 1843, the Queen, together with Prince Albert and assorted courtiers, left Buckingham Palace and made the short drive to Whitehall where the party embarked on the State Barge. For the first time in nearly two centuries, the sovereign of Great Britain would be visiting Wapping in the East End, but the precise destination was not generally known until shortly before the Queen stepped onto her shallop, as the royal barge was known, and set off downstream, propelled by her oarsmen in their scarlet livery, and accompanied by the Admiralty barge, two police barges, and a flotilla of a hundred craft or more gathered around to celebrate the expedition. The nearest modern equivalent would be Queen Elizabeth's Jubilee pageant on the Thames in the summer of 2012.

---

5 *The Times*, 27 March 1843.

Three and a half miles to the east, preparations were under way for the royal visit. Flags were hoisted on every house, the bells of the old church were ringing, and crowds gathered near the entrance of the Thames Tunnel, celebrated as one of the wonders of the modern world. By the time the Queen made her visit to the tunnel, millions of her subjects, and many thousands of curious foreigners, had already made their way to Wapping and paid a penny for the privilege of making their way under the Thames. The tunnel enjoyed great success as a tourist attraction, even if it never did fulfil its potential as a commercial highway. So little notice was given of the royal visit that Marc Brunel could not be there to receive her, though in the crowds of well-wishers, members of the nobility and elegantly dressed ladies intermingled with ragged cockneys, weather-beaten seamen and black-faced coal porters. 'The company was more numerous than select,' noted a snooty newspaper reporter. But the Queen herself almost did not make it, coming close to being run down by a steamer, as she recorded tremulously in her diary that night:

> The river was full of steamers & shipping all kinds [sic], and off Horsley Down [near the site of the future Tower Bridge], a steamer got athwart our Barge; for a second we were in great danger of being upset, & it was a dreadful moment. The Admiralty Barge, which was leading the way, chose to cross from side to another, & the steamer in consequence did not see us & came straight upon us but mercifully that Providence, which ever watches over us, again saved us. Clem screamed & I also cried out 'turn the boat', God knows I felt as if all was over. Albert turned as white as a sheet & I think all [...] were much terrified, though they did not own it at the time. He, & others, since said that we had been in great danger, & that the river was very unsafe.[6]

At half past three, guns were fired to herald the approach of the royal flotilla; the Tunnel Pier was clad with crimson cloth; people stood waiting excitedly on the quay or perched precariously on the roofs of nearby houses. Five minutes later, the Queen disembarked and was greeted with three loud cheers, which she acknowledged without a hint of inner turmoil. 'The people Received us most kindly & cheered tremendously,' she recalled, concealing a state of terrible agitation.

---

6 Queen Victoria's diaries are available online at http://www.queenvictoriasjournals. org/home.do Accessed 26 June 2015.

With the fortitude that is expected of a British monarch, she then descended the tunnel shaft. 'One goes down a long way,' she noted, '& then enters the Tunnel, which was lit by gas'. She walked almost all the way to Rotherhithe, before returning the same way as she had come. Even down in the brick-clad tunnel, there were crowds of admirers, including a handkerchief maker who cast his wares on the floor for her to walk on. 'There were many people there,' she wrote, 'who cheered us very much, which had an almost strangely awful effect in that subterranean place, where they sang 'God save the Queen'.' This may have been one of the wonders of the modern world, but the monarch found it physically uncomfortable and oppressive. 'It was excessively hot both going & returning,' she recorded. Back on the surface, she was greeted by fifty coaleys (or coalmen) from the local firm of Irving & Brown, who cheered and raised their fan-tail hats, their faces blackened by coal dust. 'God bless you Marm,' one shouted as he threw his hat into the air, 'I hope you'll come to Wapping again.'

Her Majesty smiled, but otherwise made no response. She returned without further incident to Buckingham Palace in time for dinner with the ageing Lord Melbourne, the former prime minister and relic of another age. That night in her prayers she expressed full gratitude for her preservation. She never did return to Wapping.

* * *

The tunnel had of course been intended to take road traffic, but it proved too expensive at the last to build the access ramps. So it was opened as a tourist attraction, a pedestrian thoroughfare through which the curious could pass at the cost of a penny. Some 50,000 people queued up to walk through on the first day; more than a million sightseers visited within 15 weeks of its opening, and 2 million over the first nine months. Revenues from this foot traffic amounted to nearly ten thousand pounds a year and more than covered the operating costs of the tunnel.

For a quarter of a century the structure was only open to foot passengers, who were attracted by numerous stalls and shops kept in the 63 underground arches, which turned the tunnel into a bazaar, where you could buy souvenirs, refresh yourself with tea or coffee, test your strength or blow into a machine to measure the capacity of your lungs, or study a diorama of the Battle of Waterloo or Queen Victoria's wedding – all while listening to a steam-powered organ playing 'Rule Britannia' and other popular tunes. A total of 66,358 people attended the first annual three-day tunnel fair,

and tourists came from all round the world, including a visiting tribe of
Ojibwa Indians from Canada, who were particularly attracted by shiny
medallions with a likeness of Brunel on them. In time, however, the tunnel
lost its allure as a tourist attraction and became smelly and dangerous, a
haunt of pickpockets and prostitutes. The income generated was not enough
to pay the interest on the government loan, let alone to start paying back the
principal. As the chairman of the company backing the project remarked,
'the tunnel remained an object of art, but not an investment of a profitable
nature.' It was finally closed as a public footway on the evening of Tuesday,
20 July 1869, by which time the East London Railway Company had bought
the tunnel for just over £200,000, a third of the total cost of £600,000, with
none of this going to the original backers.

The project was thus a disaster for its shareholders, but as the world's first
true subaqueous tunnel it proved to be of considerable benefit for society a
whole. 'I have often said and thought that it is the greatest engineering work
in the country,' said James Walker, the second president of the Institution
of Civil Engineers, in 1847, 'on account of its final success after so many
accidents, and the means that were taken to overcome them.' The work
helped Marc Brunel recover from the stigma of bankruptcy and win
recognition for his many achievements: he was knighted in 1841. Maudslay's
shield-tunnelling device became the template for future tunnel construction
the world over: there would be no Eurotunnel without the device designed
by Brunel and made by Maudslay. Similarly, the thousand-ton tunnel-
boring machines (TBM's) creating the 26 miles of tunnels for the Crossrail
project, cutting through the clays under London at the rate of 100 metres a
week, operate on the same principle. The original Thames Tunnel remains
in service today, connecting Rotherhithe to Wapping on the London
Overground East London Line. Weary commuters to Canary Wharf little
suspect that they are riding through one of the wonders of the nineteenth
century world, though enthusiasts can visit a modest museum on the site of
the Rotherhithe access shaft and, on occasion, descend underground for a
concert or dinner in homage to the pioneers.

CHAPTER 5

RICHARD ROBERTS AND THE IRON
MAN OF MANCHESTER

Joshua Field, Maudslay's business partner, was a great engineer, but perhaps not an imaginative man. His account of a tour of the industrial districts of the Midlands and Lancashire in the summer of 1821 contains many diagrams of machines, foundries and factories, and an account of the home near the Soho Foundry to which William Murdoch, James Watt's right hand man, had retired. The doorbell was moved not by a wire, 'but by a column of air in a pipe having a piece of glass tube at each end with Pistons fitted in, so moving the piston at the door, the piston at the bell having a knob upon it strikes the bell.' It performed perfectly well and had been in action for years, Field observed. He was impressed that the gas for heating the house was piped half a mile across fields from the factory. By the time he got to Manchester, a few weeks after that tour, Field looked up Richard Roberts, the first of the Maudslay men to leave Lambeth for Lancashire. 'He is a good workman & has good ideas with many very good tools,' Field observed, but then added, somewhat disparagingly, that his former employee's workshop was 'in a bad part of town and extremely shabby outside.' After studying Roberts's lathe with back-geared headstock, Field went for a walk around the town. 'Mr. R took us to see the Field of Peterloo,' Field noted. '[Roberts] pointed out where the Massacre took place, where he stood, and where he made his escape.' Then Field visited the Manchester gasworks, a ribbon-weaving factory and a fustian cutter at work in a garret.[1]

Field thus glossed over the Industrial Revolution's equivalent of the Tiananmen Square massacre, and he failed to record anything out of the ordinary about Manchester, on the cusp of an extraordinary expansion of

---

1 Joshua Field, 'Diary of a Tour in 1821 through the Provinces,' *TNS*, vol. 6 (1825): 25–26.

population and already a by-word for unbridled industrialization and resulting dehumanization and squalor. As for Roberts himself, Field's remarks gave no hint that his former employee would become one of the great engineers of the age. He was to be, or at least to manufacture, the Iron Man of Manchester, an epithet bestowed on one of his ingenious machines for the textile industry, but it might as well have described this modest but extraordinarily gifted mechanical engineer, himself. James Nasmyth, who arrived in Manchester in 1834, called Roberts 'one of the most capable men of his time, entitled to be regarded as one of the true pioneers of modern mechanical mechanism'. Karl Marx called Roberts's self-acting mule 'the greatest invention in modern industry,' crediting it with opening up 'a new epoch in the automatic system'.[2] A quixotic, almost tragic figure, Roberts was obsessed with invention for its own sake and, despite his success in building a major industrial company, he was to die a near pauper.

Roberts left Maudslay's for the North in 1816 – as did others, he went when there was no danger that he would be called into the army. He opened a workshop at No 15, Deansgate, in the centre of Manchester, where he advertised as a 'plain and eccentric turner,' suggesting that at first he worked in wood rather than metal, using a lathe to fashion wood into oval shapes. He was soon at the forefront of the move away from wood, leather and brass, to iron, transmitting the Maudslay manufacturing principles to the industries of the North. The cotton industry was booming and ripe for the application of greater power and mechanization, particularly as steam power supplanted water. In the 1820s, William Fairbairn had made a breakthrough in the design of textile mills by replacing cumbrous wood and cast-iron shafting systems that powered the weaving or spinning machinery with stronger and lighter wrought iron. 'All the cotton mills that I had seen were driven with large square shafts and wooden drums, some of them 4 feet in diameter,' Fairbairn recalled. 'The main shafts seldom exceeded 40 revolutions per minute; and although the machinery varied in velocity from 100 up to 3,000 revolutions, the speeds were mostly got up by a series of straps and counter drums, which not only crowded the rooms, but seriously obstructed the light where most required.' By introducing lighter shafts and smaller drums, Fairbairn tripled the speed of the transmission and reduced costs, decluttering the inside of the mills and letting in more light. As his system was applied throughout the North-West, the mills became larger and the machines faster. There was a premium on

---

2 See Chapter II of *The Poverty of Philosophy* (1847) and *Das Capital*.

speed, strength and accuracy. By 1835, three quarters of the power used in the industry came from steam.

Textiles dominated all, with some 30 per cent of the population employed in the production of cottons. The industry defined the industrial landscape of the North-West: in the latter decades of the eighteenth century, hundreds of mills were built in Manchester and the thriving industrial regions of Lancashire and the West Riding of Yorkshire. By 1840, 70 per cent of the 1,105 mills in Great Britain were in Lancashire, while about 1.5 million people were directly and indirectly dependent on employment in the production of cotton, and cotton cloth constituted about half of all British exports. Manchester, Bury, Oldham and a handful of other Lancashire towns had more than a hundred mills each, an extraordinary concentration of production. 'Never before in the history of humankind had such a confined area produced for such a vast number of consumers around the world,' observes Giorgio Riello, a historian of the cotton industry.[3]

The industry had its own language, and there is poetry to early accounts of the mills at work: '[T]he rovings have four hanks to the pound, and are spun into yarn on the throstle, as well as the mule,' wrote Andrew Ure of one modern factory in Lancashire.[4] The raw cotton fabric was first cleaned, then carded (in which tufts and knots were eliminated and the fibres laid down in parallel in the form of a ribbon or riband), then doubled or drawn out to 'complete the parallelism of the filaments', and through the process of fine roving, stretched or drawn out into even finer fibres, which would then be spun on a throstle frame or a mule to create the yarn, which would then be packed up to be shipped off to be woven, in a separate series of operations, typically at another factory. The throstle was distinct from a mule, in that it operated continuously and it made a singing or humming sound as it worked, hence the name, a dialect word for a song-thrush.

Early mills employed typically no more than 200 people, many of them women and children. In his influential *Philosophy of Manufactures*, Andrew Ure ventured that the children working 12 to 14 hour days lived healthy lives:

> I have visited many factories, both in Manchester and in the surrounding districts [...] and I never saw a single instance of corporal chastisement inflicted on a child, nor indeed did I see children in ill-humour. They

---

3 See Giorgio Riello *Cotton: The Fabric that Made the Modern World* (2013), 228, and Robert Allen, *Global Economic History*, 32.

4 Andrew Ure, *The Cotton Manufacture of Great Britain* (1861), vol. 1, 310.

seemed to be always cheerful and alert, taking pleasure in the light play of their muscles, – enjoying the mobility natural to their age. It was delightful to observe the nimbleness with which they pieced the broken ends [...] and to see them at leisure, after a few seconds' exercise of their tiny fingers, to amuse themselves in any attitude they chose, til the stretch and winding on were [...] completed. The work of these lively elves seemed to resemble a sport, in which habit gave rise to a pleasing dexterity. [...] As to exhaustion by the day's work, they evinced no trace of it on emerging from the mill in the evening; for they immediately began to skip about any neighbouring play-ground, and to commence their little amusements with the same alacrity of boys issuing from a school.[5]

Despite such blandishments, and a succession of Factory Acts other legislation designed to improve the lot of children and others working in cotton mills, conditions were harsh. De Tocqueville noted in 1835 that three quarters of the 1,500 employees at the eight-storey Redhill Street mill were women and children, working an average of 69 hours a week.

By the beginning of the century, factories were getting larger. Peter Gaskell, an early observer of the industry, attributed this precisely to the years 1801–04, when the 'application of steam as a moving power became general'.[6] A good example was the mill built for Philips and Lee, cotton spinners of Chapel Street, Salford, erected around the turn of the century to a design by Boulton and Watt. It was seven stories high and housed 700 operatives at the time Roberts came back North. It had an iron frame and was lit with gas, thanks to William Murdoch's invention. Other pioneers of larger mills were the Peel family, who between them at one point had more than a dozen factories in and around Manchester (and produced Robert Peel, the Conservative prime minister); McConnel and Kennedy, whose huge mill is still to be seen at Redhill Street on the banks of the Rochdale Canal in Ancoats; and Marshalls of Leeds and Shrewsbury. By the end of the Napoleonic Wars, John Marshall's three flax-spinning mills employed 1,200 people, spun 150,000 bundles of yarn and generated revenues of £250,000 a year.

There were specializations and interdependencies, with mills dedicated to spinning, weaving and printing of various different materials, from silk to cotton, coarse fustian and gaudily coloured printed calico for exporting back to

---

5 Ure, *Cotton Manufacture* 301.
6 Peter Gaskell, *The Manufacturing Population of England, its Moral, Social and Physical Condition* (1833), 179.

the colonies. All moved quickly, the machines appearing to chase one another in the hurried pursuit of production and profits. As the French journalist Léon Faucher later observed: 'In less than eight days, the cotton spun at Manchester, Bolton, Oldham or Ashton, is woven in the sheds of Bolton, Stalybridge or Stockport; dyed and printed at Blackburn, Chorley or Preston, and finished, measured and packed at Manchester. By this division of labour [...] the water, coal and machinery work incessantly. Execution is almost as quick as thought. Man acquires, so to speak, the power of creation, and he has only to say, "Let the fabrics exist," and they exist.'[7] 'An Englishman [...] labors three times as many hours in the course of a year, as any other European,' noted the visiting American writer Ralph Waldo Emerson. 'He works fast. Everything in England is at quick pace. They have reinforced their own productivity, by the creation of that marvelous machinery which differences this age from any other age.'[8]

When Roberts arrived in Manchester, the population of the town was on the verge of explosive growth: it had risen from 24,000 in 1773 to 70,000 in 1801 when the first formal census was conducted, to 110,000 in 1814, and would climb to 250,409 in 1841; in the decade 1821 to 1831, the population jumped by a staggering 45 per cent. These figures are for Manchester proper; add in Salford, on the other side of the stinking River Irwell, and there were some 400,000 living in a giant conurbation by mid-century, a hyper-modern urban centre to rival Shanghai or Mumbai in our own time. In 1814, the Swiss industrialist Hans Caspar Escher was amazed that in a single street in Manchester, there were more cotton spindles than in the whole of his own country: in the space of a 15-minute walk he counted 60 spinning mills. He did not stray off the main roads, but had he taken a turn into the backstreets, he would have found the cramped and stinking hovels of the poor. Manchester was the first city to experience the cramming of mills and factories into the town to be near the labour force, a process we know as urban industrialization. As one historian puts it, Manchester was 'a new kind of city in which the formation of a new kind of human world seemed to be occurring [...] widely regarded as the ur-scene, concentrated specimen, and paradigm of what [the industrial revolution] portended for good and bad.'[9] There were 1,369 steam engines and 866 waterwheels at work by the time of Queen Victoria's accession in 1837. Manchester and its environs were also at the heart of international

---

7 Leon Faucher, *Manchester in 1844: Its present condition and future prospects* (1844), 16.

8 Ralph Waldo Emerson, *English Traits* (1856); see Chapter 10, 'Wealth'.

9 Steven Marcus, *Engels, Manchester and the Working Class* (1974), 3.

trade, especially from 1830 when the railway brought down the time and cost of shipping goods to and from the giant port of Liverpool.

The town attracted superlatives of praise and denunciation. 'Manchester streets may be irregular [...] its smoke may be dense and its mud ultra muddy, but not any or all of those things can prevent the image of a great city rising before us as the very symbol of civilization, foremost in the march of improvement, a grand incarnation of progress.'[10] Alexis de Tocqueville visited in 1835, noting that in Manchester 'humanity attains its most complete development and its most brutish'. 'Manchester [is] an agglomeration the most extraordinary, the most interesting and in some respects, the most monstrous, which the progress of society has presented,' wrote the French journalist Léon Faucher.[11] Poverty, degradation and dehumanization prevailed – as described later in the century in terms of horrified fascination by novelists such as Elizabeth Gaskell, Benjamin Disraeli and Charles Dickens, and social scientists like James Kay-Shuttleworth, de Tocqueville and Friedrich Engels, whose descriptions of living conditions in working-class Manchester are lurid in the extreme. 'Hell on earth [...] everything here arouses horror and indignation,' is how Engels portrayed the Angel Meadow area. (He lived in Manchester from 1842–44, and the book was published in Germany in 1845, but not in English until 1891). The place was 'full of cellars inhabited by prostitutes, their bullies, thieves, cadgers, vagrants [and] tramps.' Houses were thrown up without forethought, and living conditions were extraordinarily squalid, especially for the 'low Irish,' unhappy wretches who lived in 'the very worst sties of filth and darkness,' little better than the hogs with whom they shared their slum accommodation. Mrs. Gaskell in her novels tiptoed around the human and animal excrement piled up in the stinking courts of central Manchester. Engels was less squeamish, making clear that the streets of the city were bespattered with filth.

Beyond the stink, there was 'anger of epic intensity [...] and a concerted movement of resistance'. [...] 'Everywhere barbarous indifference, hard egotism on one hand, and nameless misery on the other, everywhere social warfare, every man's house in a state of siege, everywhere reciprocal plundering under the protection of the law'.[12] It was a city fashioned for work and industrial production, with no concessions made to common humanity. 'It

---

10 Chambers' *Edinburgh Journal*, 1858
11 Faucher, *Manchester in 1844*. transl. 1844, 3.
12 Friederich Engels, *The Condition of the Working Class in England in 1844* (first published in German in 1845) Penguin (2009), 37.

is essentially a place of business,' wrote Dr W. Cooke Taylor following a visit in 1841, 'and amusements scarcely rank as secondary considerations. Every person who passes you in the street, has the look of thought and the step of haste.' Visitors noticed that during working hours, there were few people to be seen. 'Manchester does not present the bustle either of London or Liverpool,' wrote Faucher. 'During the greater part of the day the town is silent, and appears almost deserted. The heavily laden boats glide noiselessly along the canals, not at the feet of palaces, as in Venice, but between rows of immense factories, which divide between themselves the air, the water, and the fire. [...] You hear nothing but the breathing of the vast machines, sending forth fire and smoke through their tall chimneys.'[13]

Despite its extreme modernity, Manchester was a manor belonging to the Mosley family, administered by a so-called court leet, a hangover from the days of feudalism, and did not even become a borough until 1838. The huge struggle for parliamentary reform gave Manchester two seats, and the whole of Lancashire had just 26 seats, with 16,632 voters out of a population of 1.29 million. Only in 1846 did the family, forebears of the fascist Sir Oswald, sell their lordship of the manor.

In the absence of central civil authority, unbridled industrialization led to mass poverty and social division between the factory owners and their employees. This conflict came to a tragic head with the Peterloo Massacre of August 1819. The name was a sarcastic Mancunian tribute to the brutality of the soldiery who earlier had been viewed as heroes for their defeat of Napoleon, and who now turned their weapons on unarmed citizens. The massacre took place in Peter's Field in central Manchester, where some 80,000 operatives, working men and women from the mills of Manchester and Salford and the textile towns, converged on this open space to demonstrate in favour of political representation and better working conditions. Local militia were given an order to arrest the radical orator Henry Hunt. Drunk after hours waiting in a nearby tavern, the horsemen cut through the crowd, and 19 people, including women and at least one child were killed. Some four hundred were injured. As one eye-witness wrote:

> The cavalry were in confusion: they evidently could not, with all the weight
> of man and horse, penetrate that compact mass of human beings and their
> sabres were plied to hew a way through naked held-up hands and defenceless

---

13 Faucher, op. cit., 18.

heads; and then chopped limbs and wound-gaping skulls were seen; and groans and cries were mingled with the din of that horrid confusion. [...] Many females appeared as the crowd opened; and striplings or mere youths also were found. Their cries were piteous and heart-rending, and would, one might have supposed, have disarmed any human resentment: but here their appeals were in vain. In ten minutes from the commencement of the havoc the field was an open and almost deserted space. The sun looked down through a sultry and motionless air. The curtains and blinds of the windows within view were all closed[, ...] over the whole field were strewed caps, bonnets, hats, shawls, and shoes, and other parts of male and female dress, trampled, torn, and bloody. Several mounds of human beings still remained where they had fallen, crushed down and smothered. Some of these still groaning, others with staring eyes, were gasping for breath, and others would never breathe more. All was silent save those low sounds, and the occasional snorting and pawing of steeds.[14]

The man who commanded the yeoman cavalry on that fateful day was Hugh Hornby Birley, a leading Lancashire industrialist who will figure later in this history as James Nasmyth's first and most significant financial backer. Due to Field's woeful incuriousness, we do not know how Roberts escaped, but it tells us a great deal that he was in the crowd of protesters. His sympathies were clearly with the working women and men, including artisans and mechanists who, like himself, were trying to better themselves through hard work and education and wanted appropriate representation in the antediluvian political structures of the time.

Roberts became partner in one of the largest works in Manchester as a by-product of his mechanical genius rather than as a result of conscious entrepreneurial intent. He seems to have been acutely aware that his inventions would lead to the disempowering of the working people with whom he had a strong affinity. He never became comfortable with the role of proprietor or capitalist, and on 7 April 1824 was one of those who gathered at the Bridgewater Arms in the centre of Manchester to form an adult educational establishment for skilled craftsmen. Modelled on the pioneering work of Dr George Birkbeck, who set up the Mechanics' Institute in London in 1823, the Manchester Mechanics' Institution would spread knowledge of science to manufacturing with the dual rationale of economic benefit and moral uplift. It evolved into what is today the

---

14 Samuel Bamford, *Passages in the Life of a Radical* (1864).

University of Manchester Institute of Science and Technology. In 1825, he also joined Manchester's more overtly patrician Literary and Philosophical Society, where Lancashire's businessmen rubbed shoulders with men of science – for example John Dalton – and debated the major issues of the day, not least the campaign to secure parliamentary representation for the industrial North and the repeal of the protective Corn Laws.

Roberts's first lathe was upstairs in a bedroom, powered by a large wheel in the basement, worked by his wife. Roberts equipped himself with machine tools that he used to produce a string of inventions, starting with a machine used for cutting teeth in cast iron wheels. There was a tremendous science to the cutting of such teeth for use in waterwheels and the transmission of steam power to the textiles factories and corn mills, and his machine was credited with being the first able to cut massive industrial-scale metal gears, up to 30 inches in diameter. Then came a slotting machine and most notably a planing machine, which (he explained much later) 'effects a vast amount of work compared to what could be done with the chisel and file [and] does a great variety of things which could not otherwise have been brought into use on account of their enormous expense'. (Roberts's original machine is now in the Science Museum, together with his slide lathe.) There was also the large lathe with backgears studied by Field (these allowed the machine to be operated at a great range of speeds). 'This series of inventions introduced a precision, celerity, and cheapness into that branch of production never before known,' noted a eulogistic article in the *Illustrated London News*, decades later. 'Their economical results have extended, not only to the largest engines of ocean going ships, and to the thousand-handed machineries of the great factories of Lancashire, but down to the railway-ticket press and the sewing machine.'[15] In 1817 or 1818, Roberts designed the world's first gas meter, which was commissioned by the local authorities so they could accurately measure and charge for the gas dispensed from the Manchester gasworks. It took him a week to make an ingenious device that used a water seal to prevent the gas escaping. He could not afford to patent the invention, but even if he made no money, the work was testament to his growing reputation. By 1821, he had moved to larger premises at Newmarket Buildings, Pool Fold, and employed more than a dozen people. He described himself as a 'lathe, screw-engine, screw stock manufacturer,' supplying his machines to iron founders, machine-makers, mechanics and cotton-spinners. He also developed a sideline in making iron billiard tables

15 *Illustrated London News*, 11 June 1864.

for export to the tropics, where wood would warp in the heat. These were the shabby, overcrowded premises visited by Joshua Field: the workshop was over some stables and could only be approached by climbing up a ladder. No wonder the visiting engineering grandee from London was horrified.

Roberts made machines to produce weavers' reeds, long comb-like devices that fitted into a loom, pushing the weft (the yarn that runs across the cloth) into shape immediately after the shuttle was thrown across the loom: important, delicate components that had hitherto been made by hand. His customer was Thomas Sharp, a prominent Manchester iron merchant who owned a wharf on the canal at Oxford Street and a counting house in the centre of town. In 1822, Roberts took out a patent for a mechanized power loom, made entirely of iron, which rapidly became his company's best-selling product: by the middle of decade, he was selling 80 of these a week, or 4,000 machines a year. This was the real starting point of modern power-loom weaving and an early example of mass production. Unheralded at the time, Roberts figured out the elements of standardisation that would underpin large-scale manufacturing, first in his factory, and subsequently throughout the United Kingdom and the rest of the world. As Andrew Ure recorded:

> Where many counterparts of similar pieces enter into spinning appara-
> tus, they are all made so perfectly identical in form and size, by the self
> acting tools such as planing and key groove cutting machines, that any
> one of them will at once fit into position of any of its fellows, in the general
> frame. For these and other admirable automatic instruments, which have
> so facilitated the construction and repair of factory machines and which
> are to be found at present in all our considerable cotton mills, this country
> is under the greatest obligation to Messrs Sharp Roberts Co.[16]

This complex work brought Roberts to the attention of the town's mill owners and textiles entrepreneurs, and he was soon able to transcend his status as a mere mechanist. Thomas Sharp saw the opportunity to build a substantial business by putting his family's capital behind Roberts's exceptional mechanical skills. The Welshman was thus plucked from the obscurity of his small, shoddy workshop in a dingy part of town and installed in the Globe Works, the new company's immense factory in the centre of town, bounded to the north by Falkner Street, to the east by the Rochdale Canal, and to the west

---

16 Ure, *Philosophy of Manufactures*, 37.

by Dickinson Street. The business relationship was formalized on 31 May 1826, when Thomas Sharp and two of his brothers took Roberts into partnership to form Sharp, Roberts & Co. In due course, the firm would become a pioneer in the field of machine tools: it produced a planing machine to rival that of Joseph Clement in London, and various slotting and drilling machines invented by Roberts. It became the largest producer of textiles machinery in Manchester, and subsequently the biggest manufacturer of steam locomotives, employing more than a thousand people. Roberts would stay in partnership with Thomas Sharp until the latter's death in 1841.

To the mill owners and capitalists, the only constraint on expansion was the truculence of the skilled labour required to supervise those parts of the spinning and weaving process that could not yet be automated. Over the half century since cotton had supplanted wool as the dominant textile, the industry had seen a sequence of technological innovations that had dramatically increased productivity. James Hargreaves's spinning jenny of 1764 was followed by Richard Arkwright's water frame of 1769, which grouped spindles together to be operated by a single source of power; with the invention of the flying shuttle in 1772, John Kay quadrupled the power of the loom; and Samuel Crompton in 1779 multiplied the power of the spinning wheel 'full fifty times' with his spinning mule, so-called because it combined the features of Arkwright's machine and Hargreaves's jenny. With the advent of James Watt's steam engines, the challenge was to harness steam power to Kay's handloom and Crompton's mule – a challenge only partially met in the first two decades of the nineteenth century. Weaving was more or less fully automated, but the process of spinning was not, despite the application of much engineering ingenuity in vain attempts to solve the problem. Crompton's mule produced a superior product, but still required skilled operatives to carry out the delicate process of winding the spun yarn onto spindles. The spinner had to 'put up' or reset the mule carriage up to a thousand times while spinning one set of cops, and simultaneously coordinate three complex processes, which required considerable physical effort (the spinners walked between 7 and 12 miles during their twelve hour shifts) as well as mental and physical agility. Even in the age of machines, then, spinning required considerable skill, judgment and artistry.

At periods of high demand for woven goods, it meant that that the weavers did not have enough yarn to meet orders, and spinners could hold the industry to ransom by charging ever-higher wages for their labour. Or they could sell their labour overseas, where rival nations were keen to establish their own textile industries. This came to a head in 1824, when the spinners at Stalybridge, a mill town eight miles to the east of Manchester, went on strike for higher wages. Stalybridge had been one of the earliest towns to embrace industrialization, introducing a mill using the water frame as early as 1789, but it had also been the scene of destructive Luddite riots in 1811–12. It was later described in lurid terms by Engels: 'multitudes of courts, back lanes, and remote nooks arise out of [the] confused way of building[. ...] Add to this the shocking filth, and the repulsive effect of Stalybridge, in spite of its pretty surroundings, may be readily imagined.' In the 1820s, the spinners were in a strong position and able to inflict substantial pain on the manufacturers, as summarized by the unsympathetic Andrew Ure:

> In the factories for spinning coarse yarn for calicoes, fustians and other heavy goods, the mule-spinners have [...] abused their powers beyond endurance, domineering in the most arrogant manner [...] over their masters. High wages, instead of leading to thankfulness of temper and improvement of mind, have, in too many cases, cherished pride and supplied funds for supporting refractory spirits in strikes, wantonly inflicted upon one set of mill-owners after another [...] for the purpose of degrading them into a state of servitude.[17]

The resulting turmoil led a group of mill owners to take the unprecedented step of forming a delegation to Richard Roberts to implore him to deploy his talents to develop an automatic or self-acting mule. The masters requested him to turn his attentions to spinning. The visiting American Ralph Waldo Emerson parodied their demands: they wanted a spinner who 'would not rebel, nor mutter, nor scowl, not strike for wages, nor emigrate'. As Roberts himself told a Parliamentary subcommittee in 1851:

> I said that I knew nothing of spinning, and therefore declined it; they called again, that was, on the following Tuesday; I declined again; but before seeing me on the third Tuesday, they saw my partner, the late Mr. Thomas Sharp, and requested that he would do what he could to induce me to turn my attention to it; on the third visit which they made, I promised to make the mule self-acting.[18]

There was speculation at the time that Roberts's reticence was due to his instinctive sympathy for the spinners. The desired innovation had the

---

17 Ure, *Philosophy*, 366.
18 The House of Lords Report on the Patent Law Amendment Bills 1851.

potential to put thousands out of work and substitute the terrorism of the masters for the despotism of the striking worker. 'The mill-owners wished to have it,' reported the *Engineer*, 'and set it at work simply in terrorism, but Richard Roberts refused to be a mere tool to change the monopoly from men to masters.'[19] But the combination of pressure from his business partners, and the sheer fascination with what was mechanically difficult, meant that he took on the task. To familiarize himself with the workings of Crompton's mule, he had one installed in the factory so he could study its workings and fathom how to mechanize the process. Roberts introduced a camshaft to control the various mechanisms, and a variety of feedback mechanisms to regulate the build-up of the cone-shaped cop. Within a few months, Roberts had built the Iron Man: the first self-acting mule, completed on Saturday, 16 July 1825.

Few believed it was a coincidence that, on the very next day, there was a conflagration at the Globe Works. The buildings were supposed to be fireproof, in that they were constructed of brick arches resting on iron beams. However, there were piles of timber in the yards outside the factory and large quantities of tinder-dry wood in the joiners' and wood-turners' shops. Just after 3 a.m., a young man walking down Bond Street reported the flames, but the fire was already out of control. The Norwich Union insurance company was only 80 yards down the road, but it took them 20 vital minutes to respond to the emergency, due to a shortage of horses. On arrival, the fire brigade broke down the gates to the factory and discovered that all the piles of timber in the yard were on fire. The first task was to liberate a large dog that had been yelping piteously since the onset of the fire. Unchained, the animal demonstrated its gratitude by biting the young man who had discovered the blaze. The fire climbed from floor to floor of the factory 'with terrific rapidity,' climbing through trap doors that had been left open, soon presenting a line of flames 120 yards long. The roof and the whole of the ground floor were now ablaze, and when the iron girders used to construct the factory buckled under the heat, some of the main buildings collapsed into the canal. Astonishingly, although £8,000 in damage was done, the self-acting mule and various other valuable tools were unscathed. 'Some suspicions are reported that [the disastrous fire] was the result of an incendiary,' reported the *Manchester Guardian*, but nothing was ever proved.[20]

---

19 *Engineer* XVII (1 April 1864), 197.
20 *Manchester Guardian*, 23 July 1825.

'The entire novelty and great ingenuity of [Roberts's invention] was universally admitted,' reported Ure in his account of the cotton industry.[21] Ure noted further than the machines were sometimes called Androides, 'from the Greek term, like a man'. He was mesmerized by the complexity of the machinery's multiple operations, its to-ing and fro-ing, advances and reversals, its close mimicry of the dexterity of human spinners:

> I have stood by for hours admiring the precision with which the self-actor executes its multifarious successions and reversals of movement; and feel myself fortunate in possessing a complete series of drawings capable of conveying to any attentive student a clear conception of every lever, pulley, and wheel in this automaton of unparalleled productive power – an instrument which exhibits almost every possible variety of mechanical organization.[22]

The machine was patented in 1825 and was subject to various improvements (and £12,000 of capital investment) over the succeeding five years. Despite its great ingenuity, and the scale of the investment, Roberts's invention was not immediately commercially successful: an economic slump in 1826 put a brake on sales, and there were some wrinkles in the manufacturing process. With the doggedness so typical of Roberts, he refined the design with the introduction of a feedback control system – which in essence allowed the build-up of the conical cop to take place while varying levels of speed and intensity based on its own changing shape and volume – and in 1830 he took out a second patent for his ingenious winding quadrant.[23] The machine eventually produced 15 to 20 per cent more yarn than the mules it replaced, and the yarn was subject to less stress and strain during spinning and therefore less liable to break. There were fewer stoppages and breakages and the machines could work more quickly than mere humans. The end product, the woven cloth, was also vastly superior. By 1834, Andrew Ure calculated that the machine was used in

---

21 Andrew Ure, *The Cotton Manufacture of Great Britain*, 2 vols., 1861. Vol. 2, 154. Ure gives an extensive description of others' attempts to make the spinning process self-acting, and has a detailed account of the workings of Roberts's machine.

22 Andrew Ure, *The Philosophy of Manufactures*, 3rd ed., 1861, 368. Built around a series of feedback loops, the self-acting mule is something of a 'cybernetic' mechanism.

23 The patent was granted on 1 July 1830 for 'certain improvements in the mechanism employed to render self-acting, the machines known by the name of mule, billy, jenny, jack frame or stretching frame, and all other machines of that class, whether the said machines be used to rove, slub or spin, cotton or other fibrous substances'.

60 mills working between 300,000 and 400,000 spindles, 'each one turning off steadily within twelve working hours, four hanks of yarn on average'. The yarn was unquestionably superior to that produced by hand, and could be woven into excellent calicoes, fustians and velvets.

Eventually, the self-acting mule was 'generally admitted to have exceeded the most sanguine expectations, and which has been extensively adopted'.[24] 'The perfection of Roberts's work enabled one man, with the help of two or three boys, to work 1,600 spindles as easily as he had previously worked 300, and mills were doubled in width to accommodate the much-enlarged machines.' It was not true, however, that Roberts's machine destroyed the factory spinner, as Arkwright had eliminated domestic spinning. Some mill owners grumbled that in fact the job of overseeing the complex machines was almost as specialized as the spinning processes it had replaced; certainly, the men who tended the machines were better paid than most, and they continued to form the aristocracy of Lancashire's labour force for decades to come.[25] Despite widespread take-up in the industry, Thomas Sharp complained that his £12,000 investment was never fully recovered; by 1837, profits only amounted to £7,000, and once again Richard Roberts failed to get rich. But his machines remained standard for the industry until the last cotton-spinning mules were built in 1927, more than a century after they were first invented. As the *Engineer* wrote, the self-acting mule 'surpasses any description of mechanism known in modern manufacture'.[26]

---

24 Ure, *Cotton Manufacture*, 154.

25 'The new mules had to be operated by highly skilled operatives, whose experience and dexterity prevented the yarns from either becoming so taut they could snap or so slack that they would not wind properly.' Mokyr, op. cit., 348.

26 *Engineer*, 18 March 1864.

CHAPTER 6

CHARLES BABBAGE, JOSEPH CLEMENT AND THE MECHANIZATION OF THOUGHT

At the same time as Richard Roberts was developing his pioneering machinery in Manchester, another Maudslay veteran was working on a still more ambitious project. While Roberts's machine mimicked the actions of a skilled spinner, London-based Joseph Clement aspired to the mechanization of thought. Formerly Maudslay's chief draughtsman, Clement had entered into an unlikely partnership with the great polymath Charles Babbage (1792–1871) to build the Difference Engine, a programmable mechanical calculator and mechanical forerunner of the modern computer. Unlike the self-acting mule, this machine was never built, and its practical application was not understood for a century and a half, when computers entered the mainstream. Conceived, however, as the eccentric dream of a gentleman scientist, executed by Joseph Clement, a craftsman engineer schooled in the values and practices of Henry Maudslay, the project led the way to the mechanization and mass production of Whitworth.

Babbage was one of the outstanding minds of his generation. He attained prominence as a mathematician, astronomer, social scientist, statistician and inventor, producing 83 published papers and 6 full-length monographs, covering subjects as diverse as cryptography, life assurance, demographics, submarines, chess, barometers, theology and of course mathematics and statistics. In 1816, he was elected Fellow of the Royal Society and, in 1828, appointed Lucasian Professor of Mathematics at Cambridge, a distinguished post that he held until 1839, without giving a single lecture. He stood unsuccessfully for Parliament three times. He became an expert on railway safety, designing (somewhat improbably) the cowcatcher, the device fixed to front of locomotives to shunt

cattle out of the way and the staple of many a Western. He also designed and
built the first ophthalmoscope for examining the eye.

Babbage was abrasive, and intolerant of those not as clever as he, which was
most people. As a young man at Cambridge, he annoyed his university tutors
by refusing to study the mathematics syllabus, finding it too provincial. In his
final *viva voce* exam, he appeared to question the divinity of God: the result was
a mere pass degree when he was expected to carry off the most distinguished
academic prizes. He calculated the odds against physical resurrection from the
dead at 200,000 million to one. Later, and not without his tongue in his cheek,
he wrote to the poet, Alfred Lord Tennyson, to complain about the line: 'Every
moment dies a man, every moment one is born.' He pointed out to Tennyson
that the second half of the line in his 'The Vision of Sin', an 'otherwise beautiful
poem', implied zero population growth, which self-evidently did not reflect the
conditions of modern society, and should be corrected to: 'Every moment one
and one-sixteenth is born.' 'The actual figure,' he wrote, 'is so long I cannot get it
onto a line, but I believe the figure 1 1/16 will be sufficiently accurate for poetry.'

Thomas Carlyle thought him 'eminently unpleasant [...] with his frog
mouth and viper eyes, with his hidebound wooden irony, and the aridest
egoism looking through it'.[1] 'A mixture of craven terror and venomous-
looking vehemence,' Carlyle wrote on another occasion, 'with no chin too.'
Babbage had the last word at a dinner party where Carlyle had harangued
the distinguished company about the merits of peace and quiet. 'After dinner
Babbage, in his grimmest manner, thanked Carlyle for his very interesting
lecture on silence,' reported fellow guest Charles Darwin.[2] Others enjoyed his
company: at the height of his powers in the 1830s, his large house at 1 Dorset
Place, just off Manchester Square, was the scene of fashionable soirées attended
by Charles Dickens, Brunel and many other distinguished figures of the day.

The son of a wealthy banker, Babbage did not have to worry about
whether people liked him: after his father died in February 1827, he inherited
a fortune of £100,000. 'He has his £20,000 [annual income] snug!' as Carlyle
observed.[3] Dionysius Lardner wrote more sympathetically: '[T]here is no
position in society more enviable than that of the few who unite moderate
independence with high intellectual qualities.'[4] Moderate independence was

---

1 *Letters of Thomas Carlyle*, 24 November 1840.
2 5 February 1839, 24 November 1840; also *The Life and Letters of Charles Darwin*
[London 1887], ed., F. Darwin, 1:77.
3 5 February 1838, letter to Carlyle's brother.
4 *Edinburgh Review*, July 1834, 263.

an understatement: Babbage was seriously rich, his inheritance worth tens of millions in today's money. Liberated from the need to earn a living, he could do whatever he liked. He married on leaving Cambridge in 1814, and settled to the north of Oxford Street in Marylebone, living the life of the gentleman polymath. In addition to his serious studies, he amused himself by conducting a scientific study of the reasons for broken windowpanes in a factory. He published his own table of logarithms from 1 to 108,000 in 1829, having applied nine separate stages of checking: his numbers were considered the most accurate of the day and were in use for half a century. More peculiarly, he waged a lifelong campaign against street music, enumerating 'instruments of torture' including organs, brass bands, fiddles, harps, hurdy-gurdies, drums, bagpipes, accordions, halfpenny whistles, trumpets and the human voice in various execrable forms, including the shouting out of objects for sale, religious canting and psalm-singing. He raged against encouragers of street music, including the obvious suspects like tavern keepers, servants and children, but also ladies of doubtful virtue and, even, titled ladies, 'but these are almost invariably of recent elevation, and deficient in that taste which their sex usually possess'.

Babbage's dream was nothing short of the mechanization of human intelligence. The Difference Engine was, as Lardner explained: 'A proposition to reduce arithmetic to the dominion of mechanism – to substitute an automaton for a compositor – to throw the power of thought into wheelwork.'[5]

Babbage gave two accounts of how he came up with the idea for the machine. According to one, he was sitting in his study as a young man at Cambridge, where he was in the habit of reading tables of logarithms with the aim of spotting errors, when he drifted off to sleep, the log tables open in front of him. Abruptly, he was woken in mid reverie by a fellow student, like his contemporary Coleridge by the Man from Porlock. 'Well Babbage, what are you dreaming about?' the student asked. Jolted awake, Babbage answered: 'I am thinking that all these tables [pointing to the logarithms] might be calculated by machinery.' In the other, a few years later (in 1820) he was working on a set of astronomical calculations with his great friend, John Herschel, a laborious job that required the two men to call out sequences of numbers in the hope of finding mistakes, when he jokingly exclaimed: 'I wish to God these calculations had been executed by steam!' Much as though Babbage and his fellow scientist enjoyed checking and rechecking mathematical tables,

---

5 *Edinburgh Magazine,* July 1834, 264.

the notion of automating the process appealed. Herschel seems to have replied that this was possible and desirable.

To win public funds for the project, Babbage sought to convince successive governments that the production of accurate statistical tables was a matter of national importance. Just as the block factory was necessary to wage war at sea, or the tunnel under the Thames to stimulate trade and demonstrate Great Britain's technological superiority, mechanically produced logarithm and other tables would preserve the country's maritime dominance – through greater accuracy in navigation – and bolster industrial supremacy at a time when manufacturing was moving from bodge job to precision engineering. Babbage touched upon national pride by alluding to French Republican attempts to produce accurate tables, assembled in the 1790s by human calculators. The French had found that the more uncomprehending the human calculating engines, the better the results, as if a capacity for cogitation got in the way of the mechanical process of compiling and checking accurate tables of numbers. It was surely far better to mechanize the process completely and apply steam power. This would eliminate the errors of transcription and calculation that would inevitably creep in if the compilation of tables were left to mere humans.

The machine was based on the mathematical principle that the successive differences of values of polynomials are constants. In layman's terms, this means that it was necessary only to add or subtract these differences to perform complex calculations. These sums produced numerical tables of any kind: it was important that the machine was truly general in its operations so one could in effect program it to provide tables according to a pre-determined formula. For a mathematician, this is a relatively simple concept to grasp, even if laymen find it more complicated. But the true complexity came in designing and building a machine to give physical expression to the idea. The complete machine was never built, but it would have stood ten feet high, by ten feet across and five feet deep, comprising six axes in a row, with each axis containing eighteen wheels, five inches in diameter. The wheels had cylinders or barrels on them an inch and a half high and inscribed with the ten arithmetical characters. As Doron Swade explains: 'Each digit of a number had its own wheel, and the value of a digit was represented by the amount by which the wheel rotated. The machine automatically moved the wheels to perform repeated addition according to the method of differences. The operator set the starting values on the wheels by hand, and each cycle of the engine then produced the next result in the table.'[6]

---

6 Doron Swade, *The Cogwheel Brain: Charles Babbage and the Quest to Build the First Computer* (2000), 30.

The axes, springs and frame were made out of steel, while the wheels and other components out of an alloy of copper and tin. It would have weighed 15 tons when complete and required 25,000 separate components. Samuel Smiles reported that Babbage's designs for the intricate, whirling machinery covered 400 square feet, and Lardner said that they were 'executed with extraordinary ability and precision [...] perhaps the best specimens of mechanical drawings which have ever been executed'. The machine was conceived as having two integrated parts: the computer or calculating element, and a printer that directly captured impressions of the numerical results, using soft metal or papier-mâché 'stereotypes'. It was set to correct itself and thus avoid human error, indeed human intercession.

Conceptually, this concept was stunningly ahead of its time: Babbage was dreaming of complete automation at a time when self-acting tools were very much in their infancy. Some of the machines that were being developed at around the same time were likened to human arms or even human bodies, most obviously Richard Roberts's Iron Man, but this machine was intended as a mechanized *mind*. 'In other cases, mechanical devices have substituted machines for simpler tools or for bodily labour,' said Henry Colebrooke, president of the Astronomical Society of London, when presenting Babbage with the society's gold medal in 1824, 'but [Babbage's machine] comes in place of mental exertion: it substitutes mechanical performance for an intellectual process'.[7] All a human being had to do was enter the initial values and crank the handle to set the cogs turning: the outcome was produced entirely by the machine – a momentous development as, for the first time, the result was independent of human thinking. The *practical* challenge was equally enormous: how to make the thousands of interlocking components, many of them identical to one another. Henry Maudslay had made great advances in levels of precision, but mass production had not advanced beyond the Portsmouth block factory. The batch processing of Richard Roberts's works in Manchester was still an exception. Certainly, in the absence of precision and standardization, it was impossible to put out production to multiple suppliers in different parts of the country, as no two factories could produce the same items, even if given exactly the same instructions. So Babbage's machine was premised on new standards of accuracy, and a wholly new ordering of industrial production. A prototype was launched in London in 1822, and in the following year Babbage secured £1,500 from the government's civil

7 Cited in Simon Schaffer, 'Babbage's Intelligence: Calculating Engines and the Factory System' *Critical Enquiry* 21 (Autumn 1994), 203.

contingency fund for the construction of a larger, fully engineered machine to
be known as Difference Engine No 1.

At first, the manufacturing took place in a workshop behind Babbage's
home in Marylebone, but the project soon outgrew these premises. Babbage
himself and a team of assistants conducted extensive field research into the
workshops and factories of the rapidly industrializing country. They travelled
the length and breadth of the country to find suppliers of components for his
Engine. His insights were eventually documented in his book *On the Economy
of Machinery and Manufactures*, which became an international bestseller when
published in 1833 and shaped the future of manufacturing even as his own
grandiose project was running into the sands. Despite the extensive search,
Babbage could not find engineers able to do the job. According to Smiles, he
eventually consulted Marc Brunel, who directed him to Clement. They started
working together in 1825 or 1826, and Clement worked on the project for a
total of six years, interrupted by a period of a year when he downed tools amid
an acrimonious dispute. The project was finally abandoned in 1834.

It is not surprising that Babbage and Clement fell out: the scientific grandee
and the weaver's son with a strong dalesman's accent were worlds apart socially,
and both were bull-headed and supremely confident of their own abilities. And
while Babbage was highly educated, Clement was self-taught, indeed virtually
illiterate, if Samuel Smiles's account is to be believed. 'Heavy-browed without
any polish of manner or speech [...] he could read but little and could write with
difficulty'. At least, as Smiles conceded, Clement's head was a 'complete repertory
of inventions.' In his workshop in Prospect Place, St George's Rd., Vauxhall
(near to the Bethlehem Hospital, now the Imperial War Museum, and not far
from Maudslay's), Clement designed an Ellipsograph, a machine for drawing
ellipses, which surpassed anything than could be done by the human hand. He
made improvements to Maudslay's screw-cutting technology. He also built a
massive cast-iron planing machine, some 19 feet long, 4 ½ feet across and 6 feet
deep, mounted on slabs of stone and concrete and bolted into the fabric of the
building itself. Giant castings were manoeuvred to and fro and cut into precise
shapes, flat, square, circular or conical, the colossal scale of the base preventing
shudders and vibrations. At the time, the planing machine was so far in advance
of anything else available in London, that Clement did not go to the expense of
patenting the design, believing quite rightly that no one else in the metropolis
had the skill to build and operate a rival device: indeed, it took ten years before
there was another on such a scale in the capital. Messrs Fox of Derby and Richard
Roberts made similar machines at around the same time (1820–1821), but these
evolved independently of one another and it seems that none were aware of the

immense importance of the work: they paved the way to all-metal construction of bigger and ever more powerful machinery and were a crucial step in Great Britain's journey to becoming the workshop of the world. 'The greatest boon to constructive mechanism since the invention of the lathe,' as Prof Willis reflected at the time of the Great Exhibition.[8] Clement charged it out at the handsome rate of £10 a day. The machine was considered a phenomenon, and the Society of Arts took the unusual step of commissioning an adulatory and superbly illustrated article about it for its journal. This ingenious perfectionist was extraordinarily confident in his own abilities, and charged his customers accordingly. He surprised Isambard Kingdom Brunel with a bill for a batch of steam whistles that was six times more than the engineer was expecting to pay. When Brunel complained, Clement retorted that the whistles were six times better than the standard product. 'You ordered a first rate product,' he told his illustrious customer, 'you must be content to pay for it.'

The years passed, the money was spent, and components were turned out by the thousand. These included calculating wheels, figure wheels, sector wheels, a variety of clutches, intricate pieces of work in complicated shapes and sizes that must have been made by specially constructed machines and then finished by hand. He quickly attracted the crème de la crème of the engineering fraternity, men who could earn higher wages working on the Difference Engine than on more mundane commercial contracts, among them Joseph Whitworth. 'Mr. Whitworth was possessed of a special aptitude for the minute accuracy of detail in mechanical work which [was required of] the skilled workmen engaged on Babbage's machine.'[9]

Still, the machine was not built. While proponents of the project pointed out that it had taken James Watt 20 years and £50,000 merely to *improve* the steam engine, and that under the circumstances it was hardly surprising Babbage was taking so long and spending so much money, there was steady criticism of what was seen as a quixotic project with no results. For as long as Babbage was personally superintending the works, travelling from Marylebone to Lambeth to oversee the drawings, matters were kept under control. But then human tragedy intervened: in 1827, Babbage lost not merely his father, but his wife (in her mid-thirties) and two sons. Grief-stricken, he took himself off to the Continent for an extended period of recuperation, leaving management of the Difference

---

8 Prof Robert Willis of Cambridge University in 1852, in 'Machines and Tools, for working in metal, wood, and other materials', *Lectures on the Results of the Great Exhibition Delivered before the Society of Arts, Manufactures and Commerce* (1852), 314.

9 *Manchester City News*, 25 November 1865.

Engine in the hands of Herschel, the astronomer. Herschel was among the great scientists of the nineteenth century, but project management was not his forte. With Babbage out of the country, rumours circulated that Clement was overcharging for his services. Herschel was forced to issue a plaintive statement to the press. 'I am enabled to state from certain knowledge, that the whole amount of the sum originally advanced by the Government has been bona fide expended on the object of its destination,' he wrote to *The Times*. 'It has proved very far from sufficient to cover the expenses of the undertaking [...] to those conversant with mechanism, and who are aware of the multitude of tools to be invented and constructed where machinery of a nature entirely new is to be executed on such a large scale, and with perfect precision, this will not appear extraordinary. [...] The work continues in active and steady progress.'[10] Despite the protestations, Clement was clearly taking advantage, charging exorbitant prices, and costs ran out of control. A workman wrote to Babbage alleging that Clement 'chose to doze over the construction year after year for the purpose of making one thing after another'.

On his return to England in November 1828, Babbage tried to get the project back on track. On December 6, he wrote to the Duke of Wellington, then prime minister, pointing out that the project had cost £6,000 to date, of which all but £1,500 had come from his own pocket. A committee of distinguished scientists and engineers was appointed to assess whether more public funds should be allocated. These included Bryan Donkin, Sir John Herschel, Marc Brunel and George Rennie. They looked hard at the tools that had been constructed and at the voluminous drawings and at the bits and pieces of the machine that had been made as a result. 'Their verdict was, in effect, 'Go on – give more money; the thing must answer.' Perhaps this was not surprising, as so many of the committee members were Babbage's close personal friends: they declared that whatever the machine would do, it would do truly. In April 1829, he personally received £1,500 of public money by way of partial recompense for his outlay, and the project was in effect nationalized. Babbage also set in motion a process of arbitration designed to establish once and for all whether Clement was overcharging or not. He asked Donkin and Rennie to examine the accounts, while Clement nominated Henry Maudslay himself to be his representative. Clement refused to carry out further work while the investigation was under way. At this point, relations between all parties were amicable, Babbage declaring that he believed Clement did excellent work

10 *The Times*, 19 August 1828.

and should be properly paid; it was merely a question of clearing up what work had been done in return for how much money. One of the arbitrators said that Clement had demonstrated 'low cunning', and that his conduct had been offensive. Still, Babbage stood by the man whose existence he barely acknowledged in his extensive correspondence:

> He has been constructing the machine for me and into his head I have for several years been conveying all my ideas on the subject of the machine and he is consequently in full possession of all of them. At several periods during this interval he has been so ill as to be in a state of the greatest danger and I cannot describe to you the anxiety with which in such circumstances I have on coming within sight of his house strained my eyes to see if the windows were closed; and I, by his death, deprived of the result of years of anxious labour. Much of that labour is now fixed in drawings which it would take much time to make another person fully understand and much of the machinery is already executed still much remains in his mind ready to be produced and it is clearly to the greatest advantage to the progress of the machine that this should continue under my direction to execute it.[11]

The arbitrators found that Clement had been charging a fair amount and, as a result, his outstanding bills were settled and work resumed in May 1830. They also carried out a far-reaching investigation into what we would call the intellectual property implications of the project. Traditionally, a craftsman owned his tools: But was this the case here, when public money had been expended to develop specialist, extremely valuable, machine tools? And who owned the drawings: Babbage or Clement? And was Clement within his rights, as he insisted, to make copies of the components? These were new problems for the new industrial age. It was settled that while Babbage owned the drawings, the tools belonged to Clement, and he could use them to run off as many components as he liked.

In December 1832, Babbage took delivery at his home of a working model one-seventh the size of the full machine: made out of bronze and steel, it was 2.5 feet high, two feet wide and 2 feet deep, with three columns, each with six engraved figure wheels. It worked, and visitors were amazed and delighted to see it in action in Babbage's living room. For a time it must have seemed to Babbage, and indeed to his backers, that his dream was to be realized.

---

11 Cited in *Henry Maudslay and the Pioneers of the Machine Age* (2002), John Cantrell and Gillian Cookson, eds., 105–6.

He opened a new factory, near his house, to give him closer oversight. Yet, the two men fell out again, this time irrevocably, after Babbage insisted that Clement move from South London to Marylebone to be close to the new works. Clement submitted a bill to the government for relocation expenses; the government refused to pay. Clement threatened to lay off all the workmen on the project if he did not receive the money. When no payment was forthcoming, he was as good as his word: in March 1833, he fired his men, including Joseph Whitworth, who subsequently moved back to Manchester and opened his own business, transmitting all that he had learned with Maudslay and Clement to the booming industrial North. Babbage wrangled with Clement over ownership of drawings and parts, eventually taking delivery of the papers and some of the machinery in July of the following year. The final reckoning was that Clement had made some 12,000 of the components out the total of 25,000 required for the final machine. The Treasury had spent a total of £17, 478 14s 10d, of which the lion's share had gone to Clement, in addition to the several thousand pounds thrown at the project by Babbage himself. Gallingly, most of the components were melted down for scrap. At Babbage's insistence, the uncompleted, abandoned machine was put on display in the museum of King's College in the Strand, next door to Somerset House, where the models of the Portsmouth block-making machines were also exhibited.

Charles Babbage barely mentioned Clement by name in the extensive correspondence over the Difference Engine, and referred to him only as 'The Engineer' in his autobiography, as if the man who was giving mechanical expression to his great idea was a mere artificer following instructions rather than a valued professional in his own right. 'My right to dispose, as I will, of [my] inventions cannot be contested,' Babbage wrote in a letter to the Duke of Wellington in 1834. 'It is more sacred in its nature than any hereditary or acquired property, for they are the absolute creations of my own mind.'[12] Yet for all the tensions in the relationship, and their ultimate failure to build the Difference Engine, designing the machine and then making the parts was one of the great achievements of this phase of the Industrial Revolution: it was the most precisely built machine ever made, a pinnacle of precision killed by the politics (and the cost) of attaining such high standards.

---

12 Cited Simon Schaffer, op. cit, 214.

CHAPTER 7

THE TRUE BIRTH OF
THE RAILWAYS

On September 7, 1830, gifted Scottish mechanic William Muir left his home in Glasgow to seek a place at Maudslay's in London. Rather than walk, as many would have done at the time, the 24-year-old took the stagecoach and was delighted to secure the seat right next to the driver. Muir was planning to travel to London via Liverpool, where the railway to Manchester was about to be opened. 'What were stagecoachmen going to do when passengers were hauled by such iron horses as those he was going to Liverpool to see?', he asked the driver. This was a tactless question, as within months of the opening of this epochal railway, the stagecoach business for that 30-mile journey went into terminal decline, and as railways opened up across the rest of the country, the stagecoach would rapidly be rendered obsolete and consigned to nostalgic writings such as Dickens's *Pickwick Papers*. The proud coachman answered by wielding his whip, administering 'near but rather vicious cuts in the flanks' of his horses. They bounded off at speed, and the driver turned smugly to his passenger to enquire whether he could 'match that with your iron horses'. Muir found it politic to change the subject.[1]

Henry Maudslay died four months to the day after the Railway Age began in earnest with the opening of the Liverpool to Manchester Railway, and it is striking that he had very little to do directly with the steam train technology that was about to transform the physical and social fabric of the country. During the early decades of the nineteenth century, precision engineering and railways were running on parallel tracks, so to speak. While Maudslay concentrated on stationary steam engines, machine tools and marine engines, another breed of engineer was trying to work out how to build what we now know as a railway. With the tenacity of purpose that Maudslay devoted to

---

1 Robert Smiles's *Brief Memoir of William Muir*, published circa 1888, can be accessed on www.gracesguide.co.uk

refining his screws and designing his machinery, contemporaries were figuring out through a ceaseless cycle of trial and error (with more error than trial) how to put a steam engine on a wheeled vehicle, and how to make that vehicle travel – if not on Britain's appalling roads – on iron tracks, which until well after the Battle of Waterloo were too fragile to carry the weight of a steam train and carriages. The paths of the railway and precision engineering did eventually converge but, in the early decades, mechanical exactitude was less important than the bloody-mindedness characteristic of George Stephenson, the veritable Hengist of the railway.

There was a time when every schoolchild would have known that the very first public railway ran from Stockton to Darlington, opening in 1825 to link the Tyneside collieries with the River Tees and ultimately the North Sea. That celebrated line was the first great achievement of Stephenson, the unlettered son of a colliery fireman who went on to build the Liverpool to Manchester railway. With Joseph Locke, he subsequently designed and built the Grand Junction line, which linked Manchester to Birmingham, and then, with his son, Robert, built the London to Birmingham line, which connected with Manchester. Together with Isambard Kingdom Brunel's Great Western Railway, these pioneering efforts created the national rail network, the great wonder of the modern industrial world that proved infinitely more useful and enduring than the Thames Tunnel.

But the creation myth around Stockton to Darlington is open to challenge: the origins of the railways owe as much to suburban South-West London as the industrial north-east. A nondescript sign on the wall of the Ram Brewery in Wandsworth attests that the Surrey Iron Railway opened in 1803 and ran for nine miles from the mouth of the River Wandle to what is now East Croydon station, following broadly the route now taken by the Wimbledon to Croydon tram link. Granted, it was not a railway quite as we know it today. It was a track of cast-iron plates laid on stone sleepers, so no rails as we would recognize them. Nor were there any steam trains, which did not exist at this time, except in the experiments of a handful of crackpot engineers. The wagons for goods and passengers were pulled by horses, hence no sleepers between the lines as these would have got in the way of the animals' hooves. But there is no denying that this was the first public railway (disallowing for a moment a rival claim from South Wales), open to all to load their goods or travel as passengers for the journey from the Thames to the Surrey hinterland, and used primarily to transport horse dung to the country and country produce to London. It is only our knowledge of what came later that might tempt us to belittle this horse-drawn railway as anything other than the real thing.

The river Wandle is now unobtrusive, hidden away behind laundries and warehouses. But the landscape was in its day every bit as typical of the Industrial Revolution as the coal mines of the North-East or the mills of Lancashire. The Wandle was too shallow to be navigable, but it provided water for scores of factories along its banks and a conduit for their waste. At the time the railway was built, there were dye works, calico printing factories, oil, tanneries and flourmills, vinegar works and the 'gigantic machinery' of iron foundries. The railway brought their goods to the Thames. As one industrial tourist breathlessly observed, you could see

> the melting of iron; the casting of the fluid; the colossal powers of the welding hammer, the head of which, though a ton in weight, gives a stroke per second; the power of shears, which cut thick bars of iron like threads; the drawing out of iron hoops by means of rollers, and the boring of cannon.[2]

There were also distilleries and breweries, chiefly Young's, on the site of which beer has been brewed for centuries, longer than anywhere else in the United Kingdom. The highly polluted water from the Wandle was no longer used to brew the beer, but the river collected the discharge and spewed it into the Thames. By the middle of the century, there were around 50 beam engines at work on this stretch of the Wandle, a remarkable concentration of steam power. Two survive in situ, inside the Ram Brewery. Silent but still shiny and in perfect working order, they can be viewed by special appointment. Built round the corner by James Wentworth & Sons, one of these was in service 12 hours a day for 109 years until retired from use in 1976. At the time of writing, the brewery has been mothballed pending redevelopment, its deserted buildings and courtyards evoking the days when hundreds of men worked there and dray horses clopped on the cobbles, hauling barrels of beer by the cartload. But thanks to site manager and former master brewer John Hatch, a microbrewery is still operating here.

In the first two decades of the nineteenth century, dozens of other short lines were established across the United Kingdom. They typically linked mines to harbours or canals, conveying unspectacular cargoes of limestone, coal, gravel, timber and rock, and were financed by local capitalists and landowners. They were private commercial ventures and not open to the public. As one historian

---

2 Sir Richard Phillips, *A Morning's Walk from Richmond to Kew* (1817). See 63–129 for his account of Wandsworth.

of the era put it, before 1825, '[T]he rail had taken a purely personal and
local character. It had performed no great public good, and it had attracted
no great public notice.'[3] There was no sense that the world was on the cusp of
epochal social and technological change, but imaginative individuals began
to see the potential. One morning in the spring of 1816, for example, author
Richard Phillips set out to walk from Chelsea to Richmond. He was impressed
by the smog hovering over London – testament to the vast quantities of coal
consumed in the capital – and by the bustle of Marc Brunel's veneering plant
and his shoemaking factory. On crossing the river and arriving at the busy
town of Wandsworth, Phillips was far-sighted enough to ask whether the
fledgling railway could be extended nationwide:

> I felt renewed delight on witnessing at this place the economy of horse-
> labour on the iron rail-way. Yet a heavy sigh escaped me, as I thought of the
> inconceivable millions which have been spent [on the recent war], four or
> five of which might have been the means of extending *double lines of iron rail-
> ways* from London to Edinburgh, Glasgow, Holyhead, Milford, Falmouth,
> Yarmouth, Dover and Portsmouth! A reward of a single thousand would
> have supplied coaches, and other vehicles of various degrees of speed, with
> the best tackle for readily turning out.[4]

Well up with the latest technology, Phillips went so far as to make the link with
steam power:

> [W]e might, ere this, have witnessed our mail coaches running at the rate
> of ten miles an hour, drawn by a single horse, or impelled fifteen miles by
> Blenkinsop's steam-engine![5]

So why did it take so long to develop steam-powered railways? There had been
railways of sorts for centuries, typically in mining areas, where carts known as
chaldrons were pulled along wooden rails by horses, donkeys or even humans.
One such line ran outside the front door of the cottage at Wylam where George
Stephenson was born and brought up, hauling coal from the pits to the staithes
of Lemington further downriver on the Tyne. As we have seen, steam power was
first introduced early in the eighteenth century, with Newcomen's atmospheric

---

3 J. A. Francis, *History of the English Railway* (1851), 57.
4 Phillips, *A Morning's Walk from Richmond to Kew*.
5 Phillips, *A Morning's Walk from Richmond to Kew*.

engines and, from circa 1780, Boulton and Watt's more efficient engines became generally available. Should it not have been relatively straightforward to make the connection between the two? In fact, there were many ingenious engineers who worked independently on the problem. William Murdoch, the inventor of gas lighting who worked for Boulton and Watt, took out a patent for a steam-powered vehicle as early as 1784, but nothing came of it. Richard Trevithick, the Cornish mine captain, is now credited with inventing and operating the first ever steam-powered railway: in response to a bet between two Welsh ironmasters, he mounted a static high pressure engine on rails and on 21 February 1804 got it to run along a 9.5-mile route at Penydarren near Merthyr Tydfil, just to the north of the great South Wales coalfields. 'We put it on the tram road. It worked well and ran up the hill and down with ease,' he reported gleefully. The engine pulled five wagons full of iron ore, together with 70 men, a total weight of some 25 tons. However, this was something of a technological dead end as the rails cracked under the weight of the locomotive and its payload, and the boiler threatened to explode.[6] Thereafter, he developed steam carriages (known as Captain Trevithick's dragons) that plied their unsteady way along the roads of Cornwall before being exhibited in Euston Square, London. Trevithick's 'steam circus' closed down after two months as the engines proved too heavy for the rails.

This early generation (most of whom, like Trevithick, came from a mining background) confronted the problems of steam locomotion: how to build boilers that did not explode; how to get the correct balance between engine-power and weight of machinery; how to get the machines and their payload to move along metal rails without falling off or smashing the rails. For years, there was a widespread misapprehension that the friction between the train and the track would be too great to allow heavy loads to be pulled along at speed, and the Heath-Robinson-esque inventions of some of the pioneers were designed to avert what turned out to be in the end a non-existent problem. Thus, some early locomotives were built like mechanized horses, with fore- and hind-legs-cum stabilisers. John Blenkinsop, an engineer working for the Middleton collieries near Leeds, designed a steam train inspired by Trevithick's notion of strong steam, but propelled by toothed wheels that slotted into a tooth-rack lying alongside the main railway. Although this rack–rail mechanism would soon be superseded, his locomotive made its maiden voyage on 24 June 1812

6 See Stanley Mercer, 'Trevithick and the Merthyr Tramroad', *TNS*, vol 26 (1947–49): 89–103.

when it dragged a load of 28 tons of coal one and a half miles from the coal staithe to the top of Hunslet Moor, making the return journey in 23 minutes with 50 spectators loaded on top as well. This machine was engineered by Matthew Murray of Leeds, one of northern England's few rivals to Henry Maudslay, and counts as the first commercially successful steam locomotive on a railway; the engine, and several others like it, were in commercial service in Lancashire, Yorkshire and Tyneside until the 1830s. At around the same time, William Hedley's *Puffing Billy* was deployed to replace the horses on the railway from Wylam Colliery. George Stephenson, a careful student of both, decided to do away with the rack–rail system in the *Blücher*, his first locomotive, which ran on a private railway at Killingworth colliery on 25 July 1814. The friction problem was finally solved only in the early 1820s when John Birkinshaw, an engineer at the Bedlington ironworks, worked out how to make wrought iron rails in sturdy lengths of 15 feet. The previous rails were made of cast iron, in much more fragile strips of three or four feet long. 'These rails are so much liked in our neighbourhood, that I think in a short time they will do way with the cast iron railways,' George Stephenson wrote to his son Robert. 'They make a fine line for our engines, as there are so few joints compared with the other.'[7] Thereafter, the standard for the railway age became flanged wheels running on smooth rails.

The pioneers were motivated less by science than economics: during the Napoleonic Wars, manpower and horses were scarce and expensive, and there was a considerable economic prize to be won if steam power could be developed as a substitute. That incentive abated with the end of hostilities, as British mines and manufactories were flooded with man- and horsepower. Yet horses remained an expensive form of transport: not only did they have to be fed and tended, they had to be cleaned up after: a typical horse produced 20 to 50 pounds of manure and a gallon of urine daily. The volume of horse dung deposited on the mucky streets of pre-Victorian England ran into millions of tons a year. Robert Gordon of Northwestern University suggests that the daily amount of manure 'worked out to between 5 and 10 tonnes per urban square mile, all requiring disgusting human labour to remove'. The annual costs of maintaining a horse were as much as buying one outright.[8] The ancillary costs of horsepower were high indeed.

---

7 From Killingworth Colliery, 28 June 1821. Cited http://www.gracesguide.co.uk/Bedlington_Ironworks, accessed 31 July 2015.
8 Robert J. Gordon, 'Is Economic Growth Over?' *Centre for Economic Policy Research, Policy Insight* no. 63, September 2012, 5.

When the Stockton to Darlington railway opened in 1825, financed by public subscription and open to the public (in contrast to the private ventures of the past), its promoters had little notion that passengers would want to travel on a route hitherto served by a coach carrying 14 or 15 people a week, still less that the coaches would be steam powered. The weekly traffic increased to 500–600 people, each carriage drawn by one horse, six people sitting inside and 15 to 20 clinging to the outside, 'crowding the coach on the top, sides, or in any other part where they can get a footing'. As steam trains were introduced, trade increased, new markets for lime and lead sprang into being, and the price of coal dropped. 'An obscure fishing village was changed into a considerable seaport town'; the railway turned 'the shop-keeper into a merchant; erected an exchange; gave bread to hundreds; and conferred happiness on thousands'.[9] This was a forerunner of what was to come on a more significant scale, first with the Liverpool to Manchester railway, and thereafter with the great trunk lines. Still, it was by no means a foregone conclusion that steam power would prevail. Nervous company directors debated the relative merits of horsepower, rope haulage and steam locomotives.

Partly in order to dispel such scepticism, the backers of the proposed railway called for public trials to take place at Rainhill, nine miles from Liverpool en route to Manchester. 'Independently of the [contest] proving the power of the locomotive,' wrote a contemporary historian, 'it was calculated to remove prejudices from those who might witness the trial, and thus create a certain moral effect on its behalf throughout the country.'[10] The rules made stipulations as to the weight, cost and power of the engines, which were to be put through their paces on a dead-flat 1.5 mile course and required to complete 20 return trips to simulate the 60 miles distance from Liverpool to Manchester and back. James Nasmyth was among the many curious young mechanics who came to Rainhill before the official competition; John Kennedy, the Manchester mill owner, was one of the judges, and William Fairbairn, Joseph Locke, Charles Vignoles and other future luminaries of the railway age that was about to be born, were also present. Nasmyth witnessed George Stephenson testing the Rocket (designed by his son, Robert) at speeds of up to 30 miles an hour. When the trials began in earnest on 6 October 1829, a crowd of more than ten thousand gathered to watch, including many fashionable men and women dressed up as if for a day at the races and lining both sides of the railroad as a

---

9 Francis, op. cit., 56–57.
10 Francis, op. cit., 129.

brass band played for their amusement. One of the strong contenders, Timothy Hackworth's *Sans Pareil* suffered a cracked cylinder and had to withdraw, and the favourite, John Ericsson and John Braithwaite's *Novelty*, also failed, while the *Cycloped*, a horse-powered vehicle that had not come to terms with the steam age, was obliged to retire after the horse fell through the floor of the engine. The Rocket won through on the second day, darting along as speedily 'as a swallow cleaving the sky', the air full of sounds of wonder and praise. The locomotive travelled 70 miles, with a load of 12 tonnes 15 cwt, in less than six hours, its speed ranging from 11.5 to 12 miles per hour. 'It had also travelled without any load, at the extraordinary and almost incredible speed of 32 miles an hour!' noted a typically breathless newspaper account.[11] Robert Stephenson collected the £500 prize and subsequently built the locomotives that defined the future of the railways for 150 years. 'The trials at Rainhill of the locomotives seem to have set people railway mad,' he observed, quite correctly.[12] Here is how Maudslay's former business partner described the crucial technological developments:

Up to this time [the 1820s], [wrote Joshua Field in 1848], locomotives were of a rude and cumbrous construction, having a cylindrical boiler, with a fireplace inside, and a single longitudinal flue through it, terminating in a short chimney, upon which the whole chimney depended for its draught. From a boiler of this construction, very little steam could be obtained, and a speed of about eight miles an hour absorbed all that could be generated, although the fire was so intense, that the flame generally extended beyond the end of the chimney, which was of course frequently red hot. Such was the condition of the locomotive when the Manchester and Liverpool Railway was projected, and although locomotive power was intended to be used, the maximum speed contemplated was only 12 miles an hour; had not a new discovery been made, and a new life given to the locomotive, the Manchester and Liverpool Railway would simply have been used in conveying bales of cotton and other goods, in the transit of coal, from the collieries to the shipping ports, as other railways had previously been engaged. At this period, however, a simple but most important improvement was introduced in the system of generating the steam: this was effected, by substituting for the single central tube, a cluster of longitudinal tubes of smaller diameter, between the fire-box and

---

11 *Liverpool Times*, 20 October 1829.
12 Letter to Thomas Richardson, 17 December 1829, cited in Jeaffreson and Pole, 153.

the chimney, through which the flame traversed, thereby vastly increasing the generating surface.[13]

A further innovation was a blast pipe, which caused a powerful ascending current in the chimney, increasing the heat and intensifying the generation of steam to keep pace with the speed of the engine.

> This was the real origin of the present railway system[, Field continued], and when this combination of the tubes, and the blast pipe was tried in [Robert Stephenson's] 'Rocket' prize engine, the speed attained astonished everyone, and completely altered the character of the whole enterprise. Instead of carrying goods ... passengers only were conveyed ... and the unexpected speed obtained, and the great passenger traffic which naturally followed, made that line the model upon which the present extensive railway system has been modeled.[14]

On 15 September 1830, the day of the opening of the Liverpool to Manchester railway, George Stephenson drove the Rocket at 40 mph. Five years before, he had been ridiculed for suggesting a steam engine could travel at a quarter of that speed. The actress Fanny Kemble left a charming and intelligent account of a train ride with George Stephenson, a week or two before the formal opening:

> We were introduced to the little engine which was to drag us along the rails. She (for they make these curious little fire-horses all mares) consisted of a boiler, a stove, a small platform, a bench, and behind the bench a barrel containing enough water to prevent her being thirsty for fifteen miles[; ...] she goes upon two wheels, which are her feet, and are moved by bright steel legs called pistons; these are propelled by steam, and in proportion as more steam is applied to the upper extremities (the hip-joints I suppose) of these pistons, the faster they move the wheels[. ...] The reins, bit, and bridle of this wonderful beast is a small steel handle, which applies or withdraws the steam from its legs or pistons, so that a child might manage it. The coals, which are its oats, were under the bench.[15]

---

13 *Proceedings of the Institution of Civil Engineers*, Address of the President, 1848, 30–31.
14 *Proceedings of the Institution of Civil Engineers*, Address of the President, 1848.
15 Letter from Fanny Kemble to a friend, cited in Humphrey Jennings, *Pandaemonium 1660–1886: The Coming of the Machine Age as Seen by Contemporary Observers* (1985), 187–88.

Kemble was rather inclined to pat the snorting little steam-horse, and declared herself horribly in love with its creator.

As the historian T. S. Ashton put it: '[T]he locomotive railway was the culminating triumph of the technical revolution'. Still, early passenger trains were conceived of as steam-powered stagecoaches on iron tracks. First-class coaches were so closely modelled on the horse-drawn original that a coachman stood on the outside of the carriage, keeping watch over the passengers. This practice was abandoned on the London to Birmingham line only after a postilion was decapitated when he forgot to duck going under a bridge. In fact, the success of Stephenson's Rocket and subsequent steam engines meant that the Liverpool to Manchester line became a railway as we know it from the very first. When London to Birmingham opened a few years later, it was by no means the case that the carriages were pulled at all times by locomotives: on one particularly steep section in Camden Town, for example, they were attached by rope to fixed engines which pulled them up the hill, and then lowered down the other side by the force of gravity. As late as 1840, the London and Blackwall rope railway ran along the 3.5 mile viaduct from Fenchurch Street in the City to Blackwall, with colossal steam engines at each end driving winding drums that propelled the trains at a respectable 25 miles per hour. The pair of 400 hp engines at Fenchurch Street was manufactured by Maudslay's. The firm was also contracted to provide engines to Isambard Kingdom Brunel's ill-fated attempts to build an atmospheric railway.

If there is an air of historic inevitability about George Stephenson's status as the greatest of early railway engineers, it reflects genuine achievements as well as his personal braggadocio and the hagiography of Samuel Smiles, his extremely influential biographer. Stephenson was a mechanical savant with a combative nature who shunted aside human obstacles such as sceptical legislators and reactionary landowners, as well as natural ones such as the supposedly impassable Chat Moss, a bog lying between Liverpool and Manchester, or the hills through which he bored his tunnels. Samuel Smiles's best-selling, *Life*, first published in 1857, tended to overlook Stephenson's more unpleasant character traits: coarse manners and speech, brusqueness bordering on rudeness, a liking for alcohol and aggressive business tactics. His biographer airbrushed these defects out of his account, portraying Stephenson as a benevolent force of nature, whose rough edges were forgivable in one who embodied the Victorian spirit of self-help, and did so much to make the fabric of the modern world.

Steam-propelled road carriages were developed in tandem with railways. The French engineer Nicolas-Joseph Cugnot built his steam car as early as 1769.

William Murdoch had taken out a patent, but no practical application was found for it. Bramah, too, made a steam carriage. Decades later, James Nasmyth and his brother built one that carried eight people along the Queensferry Road outside Edinburgh. Nobody could see any practical application for the vehicle, so Nasmyth broke it up and resold the parts at a profit. Another pioneer was Goldsworthy Gurney, one of whose contraptions in 1829 made the journey from London to Bath and back. ('I never heard of any accident or injury to anyone with it,' his very proud daughter, Anna, stated in a letter to *The Times*, nearly fifty years after this adventure, 'except the fray at Melksham […] when the fair people set upon it, burnt their fingers, threw stones, and wounded poor Martyn the stoker.'[16]) David Napier, the shipbuilder, claimed in 1835 that he had spent £10,000 trying in vain to develop a steam carriage, the failure serving to prove that there was no meaningful competition to the railways. Maudslay's built a number of steam carriages, always as one-offs paid for by rich and eccentric clients, including a prototype of the omnibus that on its maiden voyage from Lambeth to Croydon carried 14 passengers and travelled at 12 miles per hour. When it reached the narrow streets of Croydon, the vehicle was making so much noise, and causing so much excitement, that the judges sitting in the Assize Courts were obliged to suspend proceedings.[17] Another customer was Walter Hancock, regarded as the most successful of the early steam-carriage inventors, whose contraption attracted the damning, and wholly correct judgment of George Stephenson himself, who said that it did not matter how well-made these vehicles were, they would never be a commercial proposition.

> I have been at Mr Hancock's place, and saw his arrangements[, said Stephenson]. I thought there was a great deal of ingenuity about it, but I told him in my opinion there was not the slightest probability of making them pay. There is no doubt of their making them go on a road, but not to make them pay, for I do not think that any experienced engineer would be concerned in them. Many ingenious gentlemen have turned their attentions to it, but if they had had much experience in keeping steam-engines in order they would not have gone into it at all. The last engine made of Mr Hancock's construction was made by Maudslay and Co., and

---

16 *The Times*, 27 December 1875. This is a feisty letter in which she claims that Stephenson pinched the original idea for the steam jet from her father – subsequently rebutted in a further letter from Samuel Smiles. Nasmyth himself claimed that he had made the same invention with his road carriage, further evidence that success has many fathers.

17 The *Observer*, 11 August 1833.

they are most excellent engine builders; it must be well done if they did it.
I do not care how well they are done; I do not see the slightest probability
of their being made to answer.[18]

William Muir finally arrived at Maudslay's on 25 April 1831, having sailed
from Ireland to London. He worked alongside James Nasmyth and Joseph
Whitworth, and was rapidly promoted to foreman.

Muir was given the unusual task of fulfilling an order for a single customer,
Admiral Lord Cochrane (1775–1860). Cochrane, the 10th Earl of Dundonald,
who had commanded the national navies of Chile, Brazil and Greece after
the fall of Napoleon, was by now a somewhat rackety figure, having retired
from his naval career and married a barmaid. The peer was preoccupied with
madcap engineering schemes and came to Maudslay's to find someone to build
a steam carriage for common roads, and according to his own design. Joseph
Maudslay, son of the founder, picked Muir as the man to give expression to
this idea. 'For a considerable time from May 25, 1833, Mr. Muir was employed
upon this carriage.' We do not know what this contraption looked like, beyond
the fact that it had two cylinders, acting directly onto the crank axles, and that
it was very heavy – about five tons, according to a satirical press notice in the
Mechanics' Magazine, which suggested that the invention had proved an 'incurable
failure'. Not so, insisted an anonymous correspondent in the subsequent issue,
for the machine made several journeys and the boiler did not explode: even if
fellow Maudslay employees were nervous of travelling aboard it. 'After some
persuasion the most talented member of the firm [Muir] was induced to ride on
it,' wrote an observer. '[H]e did not exhibit any remarkable anxiety to remain
on the vehicle, but was the first to get clear of it when it stopped[; …] the
event proved that this arose merely from a nervous apprehension of imaginary
danger, for the boiler did not burst, nor did the machine do any damage that
I know of'.[19] The invention was impressive enough for Maudslay to reward his
man with a special bonus, and for Cochrane to keep in touch with Muir when
he left Maudslay's. Muir worked next for Holtzapffel, the celebrated toolmaker
in Long Acre, Covent Garden, and then Bramahs in Pimlico, before heading
in 1840 to Manchester to join Whitworth. By that time, the centre of gravity in
British engineering had shifted to the North. After Richard Roberts, the first
to make the move was James Nasmyth.

---

18 *Mechanics' Magazine*, vol. 26, 5 November 1836, no. 691, 80.
19 *Mechanics Magazine*, vol. 22, 8 November 1834, 108.

CHAPTER 8

JAMES HALL NASMYTH – THE STEAM HAMMER AND ENTREPRENEURIAL TRIUMPH IN MANCHESTER

James Nasmyth had stayed on in Lambeth for some months after his patron's death in February 1831, helping Joshua Field build a pair of gigantic pumping engines. Despite his respect for Field, the 22-year-old Scot had 'no intention of proceeding further as an assistant or journeyman [and] intended to begin business for myself'.[1] Field supported the idea, and allowed the young man to obtain castings of a substantial turning lathe which became 'the parent of a vast progeny of descendants': the tools that he built in Edinburgh, the machines that they were used to make, and then the multitude of inventions that he would produce in his own establishments. He travelled home to Edinburgh where he established a small workshop. Using his father's foot-powered lathe, he produced first a replica of the Maudslay lathe, which he then deployed to build a planing machine, which turned out near-perfect surfaces, as we have seen something of a holy grail for early engineers. These two machine tools then created all the other tools that Nasmyth needed to set up in earnest. Nasmyth correctly foresaw that the production of tools that made other tools would become a significant industry in its own right. Moreover, Nasmyth calculated that he could at least start a business without substantial capital, 'as the risks were small and the returns were quick'. This was not the case with more capital-intensive industries like iron or cotton processing, and indeed he turned down the opportunity to go into partnership with the proprietor of an iron foundry. He reasoned, quite correctly, that he had more to give, and to gain, by concentrating on the emergent field of specialist mechanical engineering.

---

1 Nasmyth, *Autobiography*, 173.

His visit to the North-West in the autumn of 1830 had convinced him that Manchester or Liverpool were the only two places to locate his business. With the help of influential friends, he soon found premises in a former cotton factory in Dale Street, just off Piccadilly in the heart of Manchester, let out in flats for manufacturing purposes. His unit was 130 feet long by 27 feet wide, powered by a shaft that was connected with a mill down the street that had spare capacity. 'There was a powerful crab crane, or hoisting apparatus, in the upper story, and the main chains came down in front of the wide door of my workshop, so that heavy castings or cases of machinery might be lifted up or let down with the utmost ease and convenience.'[2] The landlords were Wren and Bennett, agents for Boulton and Watt in Manchester, and rent was £50 a year. When Nasmyth started the business in 1834 at the age of 25, he had his tools, and cash of just £63; he turned this into a great industrial enterprise and a fabulous fortune to match, worth £250,000 when he retired in 1856. He did this through a combination of engineering talent, commercial ruthlessness and all-round entrepreneurial resourcefulness. Also, unlike many former Maudslay men, he was superbly well-connected, immediately at home in the Manchester business community. He was recognizably of the mill-owning class, even if he did not yet possess his own mill.

While looking for premises, he had letters of introduction to some of the great figures of the day. He met William Fairbairn, the former millwright who had revolutionized the textile industry by introducing wrought-iron power systems for driving machinery – these were smaller and less cumbrous than the wooden drives they replaced, and yet much more powerful. The new technology encouraged mill owners to build bigger factories, and by the time Nasmyth met him, Fairbairn was in the business of delivering 'turnkey solutions' to meet this need, building and equipping entire factories. Nasmyth visited Richard Roberts, fellow alum of the Henry Maudslay school of engineering, and he could, no doubt, have looked up Joseph Whitworth, who had established his own business two years before. But there was always considerable antagonism between these two men, and though their paths must have frequently crossed, and they had many friends in common, they never referred to each other by name. (Whitworth's name did not even make it into the index of Nasmyth's autobiography.) This seems to have been a combination of professional rivalry and personal dislike inspired by the social divide between the privileged scion of the Scottish Enlightenment and the loom-maker's son from Stockport.

---

2 Nasmyth, *Autobiography*, 182.

On his previous visit to Manchester, the Scot had befriended Edward Tootal, one of ten sons of a wealthy Yorkshire corn merchant, who had come to Manchester around 1820 to make his fortune. By the time Nasmyth met him, he was proprietor of a prominent silk-winding empire, with two mills in the centre of Manchester. Tootal subsequently became a railway entrepreneur, influential in the creation of the London and North-Western Railway (LNWR), while the firm he founded became Tootal, Broadhurst Lee, Lancashire's dominant textiles company later in the century. A close friend of Sir Robert Peel, Tootal was typical of the thrusting, risk-taking industrial aristocracy of Manchester into which Nasmyth was immediately initiated.

As in the technology booms of our own period, there were abundant opportunities for businessmen to put their capital to work and to make quick and vast fortunes. These were the men whose capital and vision had led to the completion of the Liverpool to Manchester Railway. The mill owners all met at the Exchange on Tuesday lunchtimes between 12 to 1, where they would make deals, learn about new inventions, sort out financing for new ventures and swap gossip about markets near and far. (The Royal Exchange, now a theatre, was built much later on the site of the Cotton Exchange.) Through Tootal, Nasmyth was introduced to John Kennedy, of M'Connel and Kennedy, engineers and proprietors of the largest fine-cotton spinning mills in Manchester: these included the Old and New Mills on Henry Street and the U-shaped, eight-storey Sedgewick Mill on Union Street (now Redhill Street) in Ancoats. John Chippendale, calico printer, invited Nasmyth to his office in central Manchester on a Tuesday, and took him along to the Exchange. Through these connections, Nasmyth got to know the textile magnate Hugh Hornby Birley, who had made his first fortune in flax manufacturing and was now like a modern-day Silicon Valley venture capitalist, putting money behind local inventors and engineers in the hope of backing winners. Birley had been a force behind the Liverpool to Manchester Railway and was now investing in docks, canals and engineering – and waterproof coats. From 1824, he had been in partnership with Charles Macintosh, the industrial chemist who had patented the process for waterproofing cloth that led to the eponymous raincoat. Macintosh was a remarkable entrepreneur himself: he distilled naphtha from the tar and ammonia that was the waste product of coal gas-lighting, combining this with India Rubber to create a waterproof varnish. Later, Birley backed John George Bodmer, the talented Swiss engineer who established a factory in Manchester after a short spell working for Sharp Roberts.

Though undoubtedly a successful businessman, Birley was not necessarily a sympathetic character: as we have seen, he was the man 'who led the first fatal charge' at Peterloo. This did him no harm with members of his own class and he was at the forefront of campaigns to bring parliamentary representation to the industrial districts of the north, culminating in the Reform Act of 1832. His enormous mill on Cambridge Street on the banks of the River Medlock had 600 looms and employed 2,000 people in spinning and weaving. Adjacent to the notoriously squalid Little Ireland district, this establishment was fortified, presumably to protect his enterprise against riots and arson.

Nasmyth befriended William, Daniel and John Grant, three brothers who were proprietors of a gigantic textiles enterprise in the town of Ramsbottom and partners in a Cannon Street finance house. They had started life as humble Scottish herdsmen who in 1783 walked 300 miles down from Elgin to the west Pennines, where they would establish their business. In time, they founded schools, churches and hospitals, becoming a by-word for benevolence; hence, they were thought to be the model for Dickens's Cheeryble brothers, the beneficent capitalists transplanted to London in *Nicholas Nickleby*. They erected a monument to themselves on Walmsley Hill above Ramsbottom, a 'public thank-offering for the prosperity and success they had achieved in their new home'. When, in 1806, Sir Robert Peel retired from business to concentrate on politics, the former cattlemen bought out his calico printworks at works at Nuttall nearby. The tower was demolished in World War II (out of fear that it would be used as a navigation aid by German bombers), and the Grant brothers and their enterprises are largely forgotten. William Grant invited the young Nasmyth to dinner and found himself so impressed by his countryman's story that he told him that the enormous sum of £500 would be available on the spot any Saturday morning he cared to call in at the office, for wages, stock and so forth, *at three per cent and no security*. Nasmyth whispered his thanks and 'got a kindly squeeze of the hand in return, and a kind of wink.'[3]

This swift initiation into the heart of Lancashire capitalism guaranteed him both customers and capital in the years ahead. His timing was right: '[E]very branch of manufacture shared in the prosperity of the times'. His arrival coincided with an explosion in demand for precision-engineered machinery and machine-making tools – an explosion that could not easily be satisfied by the iron foundries and general engineering works that had grown up from Shudehill to Great Bridgewater Street, along the line of the Ashton and Rochdale canals. By this time, machinery had evolved from the hybrid of cast iron, brass, wood and leather that characterized the efforts of

---

3 Nasmyth, *Autobiography*, 186–87.

the late eighteenth century. Notwithstanding Richard Roberts's influence, manufacturing techniques were still primitive when compared to those of Maudslay and his disciples. Manchester was ready for Maudslay standards of mechanical perfection. Ingenious and energetic, Nasmyth was poised to satisfy this demand.

His first commission was an emergency order for a metallic piston for the Tootal factory, which he carried out overnight. The influential Peel family became customers, as did Edward and Ebenezer Cowper, manufacturers of specialist printing machinery who shared the same landlord as Nasmyth. The Cowper brothers made machines that were among those that revolutionized the newspaper and book-publishing industry, allowing publishers to print quickly and in immense volume. In 1827, they supplied *The Times* with a steam-powered 'multiple' machine with four cylinders that produced the then unheard-of 4,000 pages per hour. 'It is scarcely imagined by the thousands who read that paper in various corners of the globe,' wrote Charles Babbage, 'what a scene of organized activity the factory presents during the whole night, or what a quantity of talent and mechanical skill is put in action for their amusement and education.' He described admiringly the process of reporting on a speech in the House of Commons, turning the reporter's shorthand into type (the job of the compositors) and the application of steam power and ink to the type. Four men fed sheets of paper into two great rollers of the printing machine, the paper was seemingly devoured as it was sucked in and applied to the inked type, and out it came again an instant later as a completely printed page of a newspaper. 'Thus, in one hour, four thousand sheets of paper are printed on one side; and an impression of twelve thousand copies, from above 300,000 moveable pieces of metal, is produced for the public in six hours.'[4] The *Penny Magazine* sold 132,000 copies a week when launched in 1832, and Bibles were produced en masse by the Oxford University Press. Until Nasmyth provided small steam engines for the printing machines, these were worked, very strenuously, by hand. Thus did Nasmyth fuel the march of progress. In time, he also sold equipment to the Royal Mint, the Woolwich Arsenal and the De La Rue banknote company.

The local manufacturers were happy to extend credit to the Scot, who found himself overwhelmed with orders, including one for an enormous steam engine to be used to power an Irish distillery. The high-pressure 20-hp engine was bigger than his usual commissions. His men were trying to manoeuvre

---

4 Babbage, op. cit., 182–83.

the engine beam in preparation for shipping, when they dropped it and it crashed through the floor of the flat into the premises below. Nasmyth's fellow tenant was a glass-cutter who made delicate objects such as decanters, many of which were now in smithereens. Neither his neighbour, nor his landlord, was amused by the incident. It was time to move. On his earlier visit to Lancashire, Nasmyth had walked the 30 miles from Liverpool to Manchester and had spotted the ideal site for a larger factory: a six-acre plot of land bounded on one side by the Bridgewater canal, on a second by the newly opened railway, and on the third side by a good road. Five miles from the centre of Manchester, the site had superlative communications: the canal was linked with the national waterway system while the railway connected it to the rapidly growing national network and was only 20 minutes from the centre of Manchester. Nasmyth established that the plot was still available, and, rather quaintly, went to visit Squire Trafford at Trafford Park, to discuss terms. He was granted a 999-year lease at an annual rent of 1 3/4d per square yard. Bridgewater Foundry, as Nasmyth called his new enterprise, opened in August 1836. The first structures were wooden workshops, but these were soon replaced with brick, and 'the hum of the driving belts, the whirl of machinery, the sound of the hammer on the anvil, gave the place an air of busy activity'. The first order was for a massive 40 hp engine for a flax mill in Londonderry, and others came in thick and fast. 'These are indeed glorious times for engineers,' he wrote to his new business partner, Holbrook Gaskell, as the new factory was in the throes of construction. 'I never was in such a state of bustle in my life [...] such quantity of people come knocking at my little office door from morning til night.'[5]

Nasmyth described the quick transition from proprietor of a small workshop to owner of an immense commercial operation that could compete with the biggest firms in the land, doing so in faux-modest terms that insinuate it was his genius alone that brought the transformation. Gaskell and others are given little credit for their part in building the company. In fact, just as James Watt depended on the commercial expertise of Matthew Boulton, and George Stephenson would have got nowhere without the backing of rich Quaker businessmen and bankers, or Richard Roberts without Thomas Sharp, Nasmyth needed know-how and capital to expand the enterprise. Without this, he might have remained a specialist engineer, like Joseph Clement, or been condemned to grow the business painfully slowly by reinvesting profits.

---

5 Letter dated 11 July 1836 cited in Musson and Robinson, *Science and Technology in the Industrial Revolution*, 494.

For this restless and ambitious man, that would never have been enough. Incontestably, though, it was because of his engineering talent and social skills that the bankers and businessmen were prepared to set him up.

The first to back him was Hugh Birley, who had been cleared of any wrongdoing for his role at Peterloo and had subsequently thrown himself into the campaign for parliamentary reform. He and his family put in more than £40,000, a colossal sum, which financed the erection of a five-storey brick building on the site, 70 feet high and 110 x 60 feet wide. The ground floor was 20 feet high and was used as a fitting up shop. There was a separate building on the eastern side for locomotive construction, and a line of buildings with total length of 400 feet and width and height of 70 feet and 21 feet, which housed the planing and heavy turning shops and the main 130-foot long foundry. Behind these were the grinding room, the engine and boiler house, the brass foundry and the ball furnace. In addition, there was a smithy, a stable and coach house and numerous dwellings for factory workers.[6] When complete, the result was a classic industrial scene memorialized by Nasmyth's father in a frequently reproduced painting, depicting the essence of Manchester capitalism.

The Birley family withdrew half their investment after only two years, and the rest by March 1839. Their successor was Henry Garnett, a young man from an iron-trading family whose descendants would remain associated with the firm until the early years of the twentieth century. Together with George Humphrys, an American-born lawyer, Garnett provided capital investment but otherwise left Nasmyth to his own devices. For day-to-day business, the most important figure was Holbrook Gaskell, the 23-year old scion of a wealthy Liverpool trading family who had started his career with Yates and Cox, prominent nail-makers and iron merchants. He entered in partnership with Nasmyth in the autumn of 1836, at which point the name of the firm was changed to Nasmyth and Gaskell. He put several thousand pounds into the business, but more important than the capital, he brought the commercial discipline that allowed the enterprise to grow to scale. He was a modern managing director, in charge of sales and marketing, accounting, pricing and what we would call supply-chain management – ensuring the timely supply of raw materials, and at the right price. Product development was Nasmyth's department, but proper accounting ensured that the firm knew the cost of what it was making and was able to charge the right amount to achieve a profit – most unusual for the time. Detailed technical records of orders and

---

6 *Manchester Guardian*, 18 June 1845.

production were useful when there were disputes with customers or when deciding which items to produce for stock. Both Gaskell and Nasmyth were under thirty, and they brought youthful enthusiasm to the task of building a business. They were consummate hustlers, working their networks of mill and colliery owners, developing relationships with the government, sniffing out opportunities with the numerous local railway companies that were springing up at this time. 'Having been informed that you have some intention of enlarging your stock of tools we beg to introduce ourselves to your intention as extensive manufacturers of Engineering tools of every kind and the very best construction,' reads a typical speculative letter. 'We shall be glad to furnish you with our prices for any machines of this kind which you may require.' 'Understanding that the Lancaster & Preston Railway Directors are in want of a further supply of Locomotive Engines we beg to solicit your orders.'[7]

Despite commercial disputes along the way, this was a mutually beneficial relationship. Sales rose from £6,800 in 1837 to £52,690 in 1841, dropping back dramatically as recession struck in the following year, before picking up with production of the steam hammer. Gaskell was entitled to between 25 per cent and a third of the profits of the firm. He stayed in partnership with Nasmyth until June 1850, when a beam in the foundry fell on his head and he decided to retire. But he recovered and enjoyed a long and successful entrepreneurial career, entering the soda manufacturing business (building a company in Widnes that was eventually absorbed by ICI), owning coal, copper and sulphur mines and becoming a newspaper proprietor, owner of what became the *Liverpool Echo*. He was active in politics, a radical liberal supporting the Anti-Corn Law League and seeking to extend the franchise and curb the powers of the House of Lords. The novelist, Elizabeth Gaskell, married the Rev William Gaskell, Holbrook's kinsman and minister of the Unitarian Cross Street Chapel, a prominent figure in the religious and cultural life of Manchester. In *North and South*, she is thought to have modelled her portrait of John Thornton, the tough but ultimately sympathetic factory owner, in part on Nasmyth. Holbrook Gaskell had a stately home at Woolton, near Liverpool, which was crammed with works of art, including paintings by Turner, Constable and Nasmyth's brother, Patrick. Gaskell died in March 1909, aged 96, leaving a fortune twice the size of his former business partner.

Holbrook Gaskell is at least mentioned in Nasmyth's autobiography, but Nasmyth's older brother, George, was expunged from the record. George was

---

7 Quoted in J. A. Cantrell, *James Nasmyth and the Bridgewater Foundry* (1984), 42.

equal partner in the earliest days of the firm and had shared many of the youthful adventures described in the *Autobiography*, from early experiments with steam power in Edinburgh to the time with Maudslay and then the start-up in Manchester. Even if George Nasmyth was not as inventive as his brother, he was far from incompetent: he was elected to the exclusive Institution of Mechanical Engineers and had some patents to his name. In the early days of Bridgewater foundry, he attended to sales, while James stayed on site, supervising production. The brothers must have been close, but in February 1843, he left the partnership after six years, returning to London to set up as a consultant engineer. Before the steam hammer came into production, this may not have been an entirely quixotic idea, and it seems he was successful at first. But at around the time his brother retired, George took the un-entrepreneurial role of director of the newly established Patent Museum, on the comfortable but not colossal salary of £300 a year. Soon after the appointment, he deployed what remained of his inventive spirit to manipulating the books, finding ever more devious ways of extorting public money. The total amount of the defalcations was just over £400, a fraction of the fortune accumulated by his brother, but once he was found out, he was dismissed (on 18 November 1859). Having destroyed as many of the records as he could, he fled to the United States, where he died in Louisville, Kentucky, on July 2, 1862, at the age of 56, in disgrace. Thus his younger and infinitely more successful brother felt he had no choice but to leave him out of his entrepreneurial history.

During all the years that he was a hard-working businessman, James Nasmyth and his wife, Anne Elizabeth, daughter of a Yorkshire ironmaster, lived in a modest cottage called *Fireside* in the village of Barton, barely six minutes' walk from the works at Patricroft on the east bank of the canal, but still preserving bucolic charms. There were dinner parties with Lord and Lady Ellesmere, and thrilling intellectual exchanges with the likes of William Fairbairn, Eaton Hodgkinson, the mathematician and structural engineer, and Bennett Woodcroft, scientific advisor of the Commissioner of Patents. 'The evenings," reported Fairbairn of gatherings at his home, Medlock Bank, 'were most agreeably spent, mostly in philosophical and scientific discussions.'[8] Nasmyth was an active member of the Manchester Literary and Philosophical Society. He dined with the Gaskells and shared with them and other intimate friends copies of his 'Fireside Sketches', uncharacteristically sentimental drawings of fairies and children gathering sticks.

---

8 R. H. Kargon, *Science in Victorian Manchester: Energy and Expertise* (1977), 46.

A corner of his factory was dedicated to building telescopes. Astronomy was as much a passion as art, and there is genuine wonderment in Nasmyth's *Autobiography* when he describes the sight of the heavens through a telescope of his own making:

> I cannot find words to express the thoughts which the impressive grandeur of the Stars, seen in the silence of the night, suggested to me, especially when I directed my Telescope, even at random, on any portion of the clear sky, and considered that each Star of the multitude it revealed to me, was a Sun![9]

There were no children. Marriage, Nasmyth declared, was 'the most sacred of all partnerships of life,' though not so sacred that he could resist the temptation to take a mistress. In time, he had an affair with a former chambermaid and fathered a child out of wedlock, as we shall see in the final chapter. He did not fraternize with the workforce, his views of the proletariat becoming increasingly jaundiced as he grew older and richer.

In time, the foundry became an enormous enterprise, employing 300 people within three years of opening and 1,500 people at its peak in the 1850s. The site expanded from six to nearly 14 acres, including four rows of houses that provided accommodation for nearly ninety workers and their families. It was visited by grandees such as the Grand Duke of Russia, Rajah Brooke of Sarawak and Chevalier Bunsen, as well as the landowners of Cheshire and Lancashire, for example Lord Ellesmere, heir to the Bridgewater Canal and other estates, who was wont to visit the works in his private barge accompanied by liveried servants. The peer was particularly impressed by the sight at twilight of molten iron being poured into moulds for giant castings. 'The white-hot fluid was run from the melting furnace into a large ladle with one or two cross handles and levers,' Nasmyth reported, 'worked by a dozen or fifteen men. The ladle contained [up to sixteen tons] of molten iron, was transferred by a crane to a mould'. Picturesque though this delicate procedure may have been, it was also exceedingly dangerous, for if the molten metal splashed from the ladle it could set a man on fire. Nasmyth improved on this rudimentary technology with his invention of his Safety Foundry Ladle, introduced in 1838, which by means of a screw wheel and worm allowed one man to manipulate the largest ladle 'and pour out its molten contents with the most perfect ease and safety'.

---

9 Nasmyth, *Autobiography*, 311.

Nasmyth sold these ladles to the Admiralty, to the Pilkingtons glass works at St Helens and to John Penn, the Greenwich shipbuilder, and the innovation was widely adopted in British industry. Nasmyth never patented the invention, presenting this as an example of his own Cheeryble-like benevolence in making the technology available to all. 'I patented very few of my inventions,' he said. 'The others I sowed broadcast over the world of practical mechanics.'[10] But in this case at least, the decision was less an example of disinterested concern for the working man, than the result of cost-benefit analysis: it would have cost £100 to take out the patent, too much to be easily recovered from sales of this low-value item.

In its heyday in the mid-forties, the factory was offering at least 126 different products. These included: pumps for feeding boilers; small, high-pressure engines used for driving machines; wrought-iron cranes for iron foundries; hydraulic presses for making lead pipes; textile machines and machine-tools such as lathes and a grooving and planing machine. He manufactured large, high-value capital goods that required prolonged and difficult casting operations. 'Such machines are now become as much articles of current demand as files or anything else of that nature,' Nasmyth wrote to Gaskell in July 1836.[11] It was pointless waiting, as Maudslay would have done, for individual orders to come in: it was rational and good commercial practice to produce a stock of goods. In 1856, a visitor saw a 'a regiment of donkey pumps all in a line [...] marshaled columns of ambidextrous lathes and grooving machines; hydraulic presses for making lead pipes; wrought iron cranes for forges etc.'[12] Producing machines for stock like this engendered economies of scale and forced a rational ordering of the manufacturing process.

In an early letter to Gaskell, Nasmyth proposed that the buildings should be 'all in a line [...] in this way we will be able to keep all in good order'. Thus, Nasmyth constructed one of the earliest assembly lines, laid out to facilitate mass production. As a fascinated contemporary observer noted:

With a view to securing the greatest amount of convenience for the removal of heavy machinery from one department to another, the entire establishment has been laid out with this object in view; and, in order to attain it, what may be called the straight line system has been adopted, that

---

10 Nasmyth, *Autobiography*, 431.
11 Cited in A. E. Musson and Eric Robinson, *Science and Technology in the Industrial Revolution* (1969), 494.
12 Cited in Musson and Robinson, ibid., 505.

is, the various workshops are all in a line, and so placed, that the greater part of the work, as it passes from one end of the foundry to the other, receives in succession, each operation which ought to follow the preceding one, so that little carrying backward and forward, or lifting up and down, is required[. ...] By means of a railroad, laid through as well as all round the shops, any casting, however ponderous or massy, may be removed with the greatest care, rapidity, and security.

The whole of this establishment is divided into departments, over each of which a foreman, or responsible person, is placed, whose duty is not only to see that the men under his superintendence produce good work, but also to endeavour to keep pace with the productive powers of all the other departments. The departments may thus be specified: – the drawing office, where the designs are made out; and the working drawings produced [...] then come the pattern makers [...] next comes the Foundry, and the iron and brass moulders; then the forgers or smiths. The chief part of the produce of the last-named pass on to the turners and planers[. ...] Then comes the fitters and filers[; ...] in conjunction with this department is a class of men called erectors, that is, men who put together the framework, and the larger part of most machines, so that the last two departments [...] bring together and give the last touches to the objects produced by the others.[13]

Customers came to the factory, saw Nasmyth's machine-tools and assembly line in action, and placed their orders – or they ordered from catalogues. In the third edition of Buchanan's *Essays on Millwork* (1841), an early Victorian bible of engineering best practice, 21 of Nasmyth's machines were featured, compared to 33 from the rest of the British manufacturing industry put together. Products included: nut-cutting and facing machines; a large variety of turning and slide lathes; screw-cutting machinery; three different types of boring mills; and machines for slotting, paring, punching, cutting, slotting, shearing and bending boiler plates. The machines were 'self-acting', that is, they had a high degree of automation. 'The machine tools [...] did not require a skilled workman to guide or watch them,' Nasmyth gloated. 'All that was necessary to superintend them was a well-selected labourer.' 'Machine tools were found to be of much greater advantage [than working men],' he wrote.

13 Love and Barton, *Manchester as It Is* (1839), 213–19, cited in Musson and Robinson, *Science and Technology*, 495.

'They displaced hand-dexterity and muscular force. They were unfailing in their action. They could not possibly go wrong in their planing and turning, because they were regulated by self-acting arrangements. They were always ready for work, and never required a holiday.'[14] The grooving and planing machine was known as Nasmyth's steam arm, as it was a direct mechanical substitute for manual labour, belonging to what Andrew Ure called 'that class of mechanical artifices in which [...] power is made to imitate the arbitrary or voluntary motions of living beings.'[15] It could be managed by an intelligent boy and was used for planing 'small surfaces and the detailed parts of machines', that is, for putting the finishing touches to a machine. As Nasmyth boasted, the work was turned out 'in so unerring and perfect a manner as not only to rival the hand work of the most skilful mechanic, but also at such a reduced cost as to place the most active hand workmen far into the background'.[16]

* * *

In his own time, Nasmyth was best known for the steam hammer which he took credit for inventing, a device that was able to bring huge amounts of power to bear with great precision of control: Nasmyth delighted in demonstrating how with one delicate movement of a 2.5 ton hammer, he could crack the top of an eggshell in a wine-glass, without harming the glass; with another, he could press enormous slabs of iron and steel used in constructing the giant bridges and other huge structures of the mid-century. Like his own steam arm or Richard Roberts's self-acting mule, the machine took on the characteristics of a human being:

> by the more or less rapid manner in which the attendant allowed the steam to enter or escape from the cylinder, any required number or intensity of blows could be delivered. Their succession might be modified in an instant; the hammer might be arrested or suspended according to the requirements of the work. The workman might thus, as it were, think in blows.[17]

For Elizabeth Gaskell, the magnificent power of the steam hammer, combined with its superb delicacy and responsiveness, recalled 'some of the stories of

---

14 Nasmyth, *Autobiography*, 295.
15 Andrew Ure, *A Dictionary of Arts, Manufactures and Mines* (1842), 83.
16 Nasmyth, *Autobiography*, 403.
17 Nasmyth, *Autobiography*, 233.

FIGURE 5 Drawing of Nasmyth's steam hammer with wine bottle, by James Nasmyth (Courtesy of Institution of Mechanical Engineers, London)

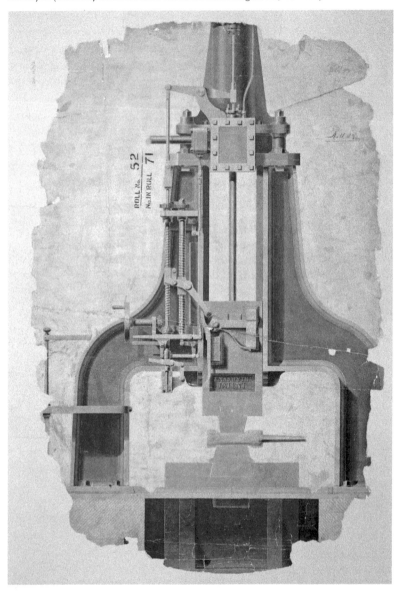

subservient genii in the Arabian Nights – one moment stretching from earth to sky and filling all the width of the horizon, at the next obediently compressed into a vase small enough to be borne in the hand of a child'.[18] Between 1843 and 1856, the factory turned out 493 hammers of different shapes and sizes, generating the fabulous sum of £286,000 in sales, nearly 40 per cent of the total for the firm. The steam hammer became the pre-eminent machine tool, that category of machines used to build other machines, without which great expansion in British manufactures could not have taken place. It remained in production until 1939, by which time the firm had produced 1,388 hammers (Figure 5).

Though ingenious engineers of the past had patented steam hammers (including James Watt), those in use by the 1840s were modelled on traditional hand hammers. Due to the angle of attack, these were too close to the object to be struck, and thus lacked any real power. The answer was to lift the hammer well above the object on the anvil. Steam power raised the hammer block, and in the final, commercially successful version of the hammer, controlled and further boosted the power of its fall beyond that provided by gravity alone. In one massive machine in use at the Woolwich Arsenal later in the century, for example, the hammer block weighed 39 tons, but with the introduction of a blast of steam above the piston, the force of the blow increased to 90 tons.

The first order for the machine had come in from Messrs Hird, Dawson and Hardy, proprietors of the Low Moor Ironworks in Bradford, Yorkshire, one of the country's most prominent producers of wrought iron. On visiting the foundry to inspect the work constructed by Nasmyth, they 'actually countermanded the order they had given, unless a better or self-acting motion could be fitted'. The prototype had none of the delicacy of action that later came to be associated with the steam hammer, none of the capacity to crack eggshells and leave glass unscathed: it was a crude, blundering sledgehammer of a tool that required the hammer block to be lifted up with the help of brute human force between each blow.[19] It took the handiwork of Robert Wilson, then the manager of the works, to refine it to the pitch of sensitivity that so impressed the Victorian public, and that Nasmyth was later so deft in displaying. Again, Nasmyth was away on business, and Gaskell asked Wilson

---

18 Elizabeth Gaskell. *North and South* (1854–55), Penguin edition, 94.

19 According to Nasmyth's original, crude conception, the hammer block, suspended from a piston in a cylinder, was raised by admitting steam to the underside of a piston. The operator admitted the steam using a lever that worked a slide valve, requiring considerable effort because of the force of the steam pressing on the valve. Thanks to John Ditchfield for this elucidation.

to address the customers' requirements as a matter of urgency. They wanted to be able to adjust the height to which the hammer could be made to rise, in order to have complete command of the power of the blow, and insisted that it should rise the instant after it struck the red hot object on the anvil, to save time and reduce the risk that the cold metal of the hammer would lower the temperature of the forging. A week after taking the matter in hand, Wilson had completed the designs and forged, turned, planed and fitted the parts to a small hammer that answered all the conditions required by the customer.

The steam hammer rapidly became Nasmyth's biggest selling product, his customers ranging from small foundries to renowned engineering companies such as his own alma mater, Maudslay Sons & Field (who bought three), John Penn of Greenwich, Boulton and Watt, and G. & J. Rennie, the Royal Naval dockyards (who bought 11 in the early years of production), railway companies such as Sharp, Roberts itself, Robert Stephenson's Newcastle operations, and the Vulcan factory at Newton-le-Willows outside Manchester. 'We are quite throng making them for all quarters,' wrote Nasmyth to Daniel Gooch of the Great Western Railway, who was considering buying one for the GWR's works at Swindon. 'The demand is rising rapidly, already we are at No 22 having made all sizes from 5 ton in the hammer block down to 1 cwt[; …] it is now quite self-acting and thumps away from 220 blows a minute to any slow rate and intensity.'[20] Another customer was Sir John Guest's ironworks at Dowlais in South Wales. Dowlais took delivery of a monster machine in April 1846, with a mass of some thirty-six tons in total. This was reputedly the largest casting in the world at the time, the hammer 'tup' or block of cast-iron, which gave the blow to the iron on the anvil, itself being six tons in weight. This machine could do the usual tricks with an egg and a wineglass, but its value here was in being able to administer six or seven gigantic blows to the piles of iron blooms used to make railway tracks … 'so as to thoroughly weld them into one solid mass ere they go to the rolls, to be extended into the finished rail'. The resulting block was bashed free of air, slag, cinders and other impurities and led to a vastly superior end product.[21] This was at the height of railway mania, but even following the crisis of 1847, orders held up. 'We are like to be smothered with Steam Hammers at the works,' Nasmyth wrote in 1853. 'We never were so busy with them before and what is most gratifying is that most of them are for folks […] who have had them from us before, each coming back for more and more and bigger and bigger ones.'[22]

---

20 Nasmyth to Gooch, 22 July 1844, cited in Musson and Robinson, 502.

21 *Stirling Observer*, 16 April 1846.

22 Nasmyth to Loch, 4 February 1853, cited in Musson and Robinson, 502.

'The steam hammer is one of the most remarkable inventions of this wonderfully inventive age,' wrote one contemporary observer. 'Without it those stupendous masses of wrought iron necessary for the construction of our ocean steamers, our ironclads, and our big guns could never have been manufactured.'[23] The tool delivered incalculable benefits to mid-Victorian manufacturing. It allowed bigger forgings to be made than ever before, at a fraction of the price: truly a revolutionary technology. Take the simple example of forging a ship's anchor. Before the steam hammer, 'the manufacture of a shaft of 15 or 20 cwt required the concentrated exertions of a large establishment, and its successful execution was regarded as a great triumph of skill'.[24] By the 1860s, forgings of 20 or 30 tons were 'things of almost everyday occurrence.' Steam hammers reduced the cost of making anchors by at least 50 per cent, while the quality of the forging was improved. The sheer size, quality and strength of the end product allowed Victorian engineers to think and operate on a larger scale than hitherto. 'In May 1874, the Emperor of Russia was shown a hammer that weighed an extraordinary 2,000 tons (including anvil and housing) which took 20 blows to bash the trunnion coil for a 38-ton gun into shape. 'Bang! Bang! Bang! goes Titan,' wrote the excitable reporter from the *Daily Telegraph*, 'and at every awful smack the red lump takes a new and denser shape, while, in spite of fifty feet of foundation, the solid earth shakes to the shock, and the air above it trembles as if the breath of distant cannonades were passing.'[25] The hammer exhibited at the Science Museum is a modest example when compared to the big beasts deployed in the armaments, shipbuilding and steel industries.

The steam hammer was adapted to driving piles, where it could apply 60 thunderous blows in a minute. These machines delivered 80 blows per minute and were capable of sinking a single, massive pile (e.g., a Baltic pine, 18 inches square and 70 feet long) into the ground in four and a half minutes, compared to the 12 hours it could take using manual labour. This tool was used in some of the massive civil engineering projects of the Victorian period, for example the dock extensions at Chatham and Devonport, the new docks at Birkenhead on the Mersey, and Robert Stephenson's High Level Bridge at Newcastle-upon-Tyne. 'Its operation has been remarkably successful,' Stephenson noted, clearly a happy customer. 'Piles have been driven with great economy and remarkable dispatch, where the ordinary methods would have entirely failed. I consider

---

23 *Hants Chronicle*, 15 September 1866.
24 T. S. Rowlandson, *History of the Steam Hammer* (1864) 7.
25 *Daily Telegraph*, 21 May 1874.

this Machine to be one of the most valuable and important auxiliaries which have recently been invented for the construction of engineering works.'[26]

Nasmyth invented many other machines, including the spiral wire drive for small drills (which evolved into the dentist's drill) and a hydraulic punching machine. His autobiography contains a long list of patents filed. 'Each new tool that I constructed had some feature of novelty about it,' he bragged, not without justification. Unlike Maudslay or Roberts, he did not pursue invention for its own sake: he was unashamedly interested in making money. Nasmyth was an outstanding engineer, always on the lookout for better or quicker ways of making things, but was a greater entrepreneur, harnessing human and financial resources, identifying the commercial potential of his and others' inventions, and marketing his products with ceaseless energy and a sure grasp of public relations, never reluctant to pose for a photograph of the steam hammer at work. He was a complex character, generous to his friends and to the memory of his mentor, Maudslay, but capable of duplicity, self-delusion and a penny-pinching meanness. He proved himself skilled in adapting to the fluctuations of a market conditioned by the frequent booms and busts of the early Victorian period. He and other Maudslay men benefited from, and helped create, the transport revolution of the 1840s.

---

26 Cited in Cantrell, op. cit., 161.

CHAPTER 9

THE MAUDSLAY MEN AND THE TRANSPORT REVOLUTION

On the evening of 23 December 1837, the engineers of the London and Birmingham Railway hosted a dinner at the Dun Cow Inn in Dunchurch, a small town set astride the great coaching routes linking the South to the North of the country, and East to West. Here, Mr. Pickwick had journeyed on his knockabout travels, but the age of Pickwick and of coach and horses was about to come to a definitive end. The dinner was given in honour of Robert Stephenson, chief engineer for the L&BR, the first main-line railway, which would open in September of the following year. The occasion was the completion of the 7,300-foot Kilsby Tunnel on the border of Warwickshire and Northamptonshire, a triumph of civil engineering and brute force. It had taken three years to complete and cost three times the original estimate of £100,000, a stupendous sum (some journalists drew parallels with the improvidence of the Thames Tunnel), chiefly because the surveyors failed to spot the presence of quicksand through which the tunnel had to be dug. Some 1,250 navvies had done the work, ripping out the fabric of the hill with pickaxes and shovels, and many had died along the way; they became a by-word for thievery, cock-fighting and drinking, their loutish behaviour the horror of more respectable citizens. 'Sheds of rude and unstable construction rose on the hill above the tunnel,' reported one contemporary with horrified fascination, 'and in them a navvy could obtain at a high rent the sixteenth part of a bedroom. Frequently one room containing four beds was occupied by eight day and eight night workmen, who slept two in a bed, and shifted their tenancies like the heroes of a well-known farce.'[1] They laid 36 million bricks and sank two gigantic

---

1 John Cordy Jeaffreson and William Pole, *Life of Robert Stephenson* (1864), vol. 1, 201.

ventilation shafts, and, with the help of steam pumps, drained the sands so the tunnel could be completed.

As the 60 revellers knew, the tunnel they were celebrating only needed to be built because of the refusal of the residents of Northampton to allow the railway to come to their city. This was a crass mistake, ensuring that the town remained a backwater while others nearby would thrive – not least Rugby, which was to become a railway entrepot and manufacturing centre. Northampton was eventually forced into the humiliating position of petitioning for a branch line to be built to connect it to the main line. A similar fate befell Stamford in Lincolnshire, where Lord Burghley insisted the railways could not travel anywhere near his estates, with the result that Peterborough flourished as a railway town as Stamford declined into genteel provincial obscurity. Robert Stephenson and his father, George, arrived at the inn at 5:30 p.m. The dinner started around 8 p.m., and the last guests did not depart until 12 hours later. It was, as L. T. C. Rolt described it as an orgy, albeit of a 'polite and orderly variety'. If there was going to be a Valhalla for Victorian engineers, this is what it would look like.

After dinner, the chairman of the L&BR, Francis Foster, presented an enormous silver tureen to Robert, and then raised a toast to the man who by now was too overwhelmed by emotion (and one suspects, alcohol) to give more than a perfunctory reply. Then a toast was raised to George. The railway pioneer got to his feet and told the story of how he prevailed against another apparently unconquerable natural obstacle, namely Chat Moss on the route between Liverpool and Manchester. For those in the room, this was a familiar story, approaching the status of a foundation myth: how he foresaw that if a ship could float on water, a railway could float on a bog; how he and his workers spent weeks pouring sand and gravel into the swamp in the hope of building a platform for the rails, with no visible progress; how the doubters had given up hope and would have withdrawn their financial backing for the entire project, but they had already committed too much and were compelled to put their faith in the rude collier's son; and how, finally, he had won through, against all the odds. It was a wonderful speech, prompting loud cheering, and no one much minded that it was not wholly true: Stephenson himself had delegated the effort on Chat Moss to Charles Vignoles (1793–1875), a lieutenant who had done most of the work, and had later been cast out of Stephenson's inner circle. Robert Stephenson withdrew from the dinner at the Dun Cow at around 2 a.m. His father had more stamina and stayed until 4 a.m., clearly benefiting from the adulation: it was thought that he looked at least half a dozen years younger than usual. 'There is the making of a hundred railways in him yet,' observed the *Railway Times*. The last stragglers left at 8 in the morning.

'Suddenly,' wrote Benjamin Disraeli of this time in his 1880 novel *Endymion*, 'there was a general feeling in the country, that its capital should be invested in railways; that the whole surface of the land should be transformed, and covered, as by a network, with these mighty means of communication.' As we have seen, the roots of the railway system reach back to early years of the century, but the opening up of the national network coincided with the beginning of the Victorian Age. The Grand Junction, linking the pioneering Liverpool to Manchester Railway with Birmingham, was a 78-mile stretch of track that passed under, or over, 150 bridges, through two tunnels, 100 excavations and embankments, and required 109,000 rails in 436,000 chairs supported on stone sleepers. This stretch opened on 4 July 1837, two weeks after Queen Victoria came to the throne on the death of her uncle, King William IV. Four days before her coronation at Westminster Abbey in June 1838, the first train ran from Euston to Birmingham, thus completing the 209-mile route from London to Liverpool, an unprecedented achievement given the sheer scale of the geo-technological challenges that had to be overcome. The first officially timetabled train left from Euston to Birmingham at 7:15 a.m. on 17 September 1838, arriving just after midday, the shortest time the journey had ever taken. There was a rush of railway building: trunk lines were connected between London and the ports of the south coast – Southampton (in May 1840); Brighton (September 1841); Dover (February 1844) – while Brunel's Great Western Railway reached Bristol in June 1841 and was extended to Exeter on 1 May 1844, reaching its final destination, Plymouth, a year later. Anyone who travels by train from Paddington Station to Bristol today can witness what Samuel Smiles calls Brunel's 'boldness of conception and design' as manifested in its elegant viaducts, bridges and tunnels as the line sweeps westwards from London. The following month, trains linked London to Gateshead, which meant for the first time the London morning newspapers could be read on the same day in Newcastle. From the summer of 1848, trains travelled from London to Edinburgh and Glasgow in around 12 hours.

In the period 1830–50, the railway industry was like financial services in the decades leading to the crash of 2008: it had an overwhelming impact on the nation, and it sucked in talent from all quarters. Just as in our times, numerate graduates find powerful financial and other incentives to work on Wall Street or the City of London, early Victorians with a halfway scientific disposition started their careers in railways. Herbert Spencer (1820–1903), the philosopher and disseminator of Charles Darwin's views (he coined the phrase 'the survival of the fittest' and was nicknamed Darwin's bulldog), worked in the 1830s as a civil engineer on the London to Birmingham line. He left an uncharacteristically light-hearted account of his youthful jaunts, for example sitting on a stone slab and sliding uncontrollably along a downhill stretch of the line. Another was Alfred Russel Wallace (1823–1913), the co-inventor with Darwin of the theory of natural selection, who made a gigantic career for himself as biologist, scientist and explorer. As a young man, he worked for three years surveying railways in south Wales and had an ambivalent attitude to them after his brother caught a chill and died following a journey in third class, where the passengers travelled like cattle, exposed to the elements in the worst of weather.

Beyond the Stephensons and Isambard Kingdom Brunel, there were others whom the Victorians acknowledged as great railway engineers. Joseph Locke (1805–1860), for example, the son of a friend of George Stephenson, was responsible for the Grand Junction. He was celebrated for delivering his projects on time and on budget. The cost per mile for this epic project ended up at £18,846 compared to an estimate of £17,000. He was equally reliable when building tunnels. His Woodhead Tunnel under the Pennines was three miles long and took five years to complete; when the two teams of engineers reached the middle, the 'driftways met with less than three inches of error'. Conditions during construction were exceedingly tough, and 26 labourers died, calculated (somewhat histrionically) as a greater proportion of those employed than English soldiers killed at the Battle of Waterloo. Between them, the triumvirate of Brunel, Stephenson and Locke constructed most of the railways built in Britain in the middle years of the nineteenth century. In Locke's mould was Thomas Brassey (1805–70), who built 3,000 miles of railway in continental Europe (including the first substantial railway tunnel in Switzerland) and a further 1,550 miles outside Europe, including the railway linking Toronto with Quebec in Canada.

Less well known is George Bidder (1806–78), the son of a mason in the Devonshire village of Moretonhampstead who, from infancy, displayed extraordinary capabilities as a mathematical savant. As a child, he astonished his father by completing mental calculations that were beyond most adults: one story has him lying in bed listening to his parents' vain attempts to figure out the value of the family pig once sent to the butchers. He became so frustrated that he shouted the answer downstairs. He would sit in the blacksmith's shop and perform sums for a penny or two. After the local schoolmaster had corroborated his powers, George's father started showing him at local fairs, as if he were a performing bear or prize pig. By the age of eight, he toured the principal towns of England and was exhibited as the Mental Calculator or the Wonderful Calculating Boy or the Calculating Phenomenon of England. Despite never being instructed in mathematics, he could solve difficult arithmetical questions without hesitation or showing any signs of mental exertion. He was able to calculate roots and cubes before he was introduced to the concepts, and on becoming familiar with logarithms and compound interest, he developed mental methods to calculate them. In April 1815, George was presented by the Bishop of Salisbury to Queen Charlotte and the royal princesses. Their questions are typical of those directed at the boy in these public displays:

Q: What number multiplied by itself will produce 36,372,951?
A: 6031 (answered in 80 seconds)
Q: How many minutes in 40 years?
A: 25,754,400, answered in 2 seconds
Q: Multiply 4,698 by 4698.
A: 22,071,204
Q: What is the cube root of 12167?
A: 23, given instantly
Q: Suppose the distance from London to Glasgow be 405 miles, how many inches are contained in the same?
A: 25,660,800, given in 15 seconds

Bidder's talents brought him to the attention of a group of benefactors, and he was given a private education, though he discharged himself from grammar school in Camberwell after six months, as he was making too much money from his public displays. In time, he became a bosom friend of Robert Stephenson and, from 1835, his business partner. They worked together on many of the great engineering projects of the day, including the Great Eastern Railway. He was a pioneer of the electric telegraph, first tested at Euston and subsequently introduced on the London and Blackwall line. He was also active in the movement to improve public sanitation, working with Joshua Field to clean up the sewage-clogged Deptford Creek on the Thames in southeast London. He ended his life a rich man, with a country house at Ravensbury on the banks of the River Wandle. His powers of mental calculation remained undiminished to the end of his life.

There is no belittling the power of Bidder's intellect, nor the force of George Stephenson's character and the all-round talent of his son Robert, but building railways was not at the cutting edge of precision engineering, at least not in the early years. Of course, it was a work of engineering genius to design and build the

Rocket and its successors. But the still-greater achievement of George and Robert Stephenson was to see 'a market in contemplation', as the modern management thinker Clayton Christensen has described that most rare entrepreneurial ability: to visualise and to will into reality an entire industry that does not yet exist and which none but you can imagine. More than mechanical inspiration, railway building and railway promotion required brute force, iron determination, political nous and connections and, by the 1840s, financial as much as civil or mechanical engineering. In his *Life of George Stephenson*, Smiles was careful to play down the degree of interdependence between Stephenson and George Hudson (1800–71), the dubious Yorkshire financier who was known as the 'Railway King'. In the boom years of the 1840s, multiple gigantic capital projects needed financing. Manipulating the stock market was as important as making the perfectly machined components required to build a steam locomotive together with its tender and carriages and indeed the rails it ran upon. In fact, as Smiles pointed out in early editions of the biography, railways ran the risk of bringing the entire engineering profession into disrepute. 'The boundless speculation of course gave abundant employment to the engineers,' he wrote. 'They were found ready to attach their names to the most daring and foolish projects, – railways through hills, across arms of the sea, over or under great rivers, spanning valleys at great heights or boring their way under ground, across barren moor, along precipices, over bogs, and through miles of London streets[….] No scheme was so mad that it did not find an engineer, so called, ready to endorse it, and give it currency[….] A thousand guineas was the price charged by one gentleman for the use of his name; and fortunate were the solicitors considered who succeeded in bagging an engineer of reputation for their prospectus.'[2]

If ethical standards were poor, so was the quality of the mechanical engineering, at least in the 1830s when the unpredicted success of the Liverpool to Manchester railway triggered sudden demand for machines and sophisticated machine-produced components in unprecedented scale. It was a standing joke, among Manchester engineers at any rate, that Robert Stephenson's factory in Newcastle was still operating in the age of hammer and chisel, at a time when the Maudslay men had mechanized all. (The partnership agreement for Stephenson's factory from June 1823 bears this out. Equipment was limited to a pair of smiths' bellows, anvils vices and three lathes; even 15 years later, the factory had a very limited array of machine tools.)[3] Certainly, the design

2 Cited in Robert Angus Buchanan, *The Engineers: A History of the Engineering Profession in Britain* (1989), 69.
3 Cited in The Rev Dr Richard Hills, *The Life and Inventions of Richard Roberts* (2002), 194.

of the Rocket was soon superseded, proving inadequate to the volume of the traffic on the line and the expectations of the railway-besotted public. 'When the Liverpool and Manchester railway was established,' noted one observer, 'it was made one of the stipulations, at the celebrated contest [the Rainhill trials], that none of the engines should weigh more than five tons, and that the rate of travelling should not be less than ten miles an hour. We now find, the very engine for which the premium was obtained, discarded as useless, and doomed to drag coal along a private railway, and engines employed upon that railway weighing upwards of twelve tons, while the public are complaining when the rate of travelling is less than twenty miles an hour.'[4] There were multiple defects with the early engines: the tubes were made of copper rather than brass, and were easily damaged; the wheels had wooden spokes in cast-iron naves, clearly inadequate to greater speed and weights; the ability of the wheels to turn was constrained, and the engines were not powerful enough (operating at a mere 50 lbs per square inch) for the new demands. Overall, running was unsteady, the cost of fuel consumed was exorbitant, and the technology in need of a fundamental overhaul. From circa 1828, Robert Stephenson had been working on a 'systematic development programme to improve the dynamic and thermodynamic characteristics of locomotive design [seeking] to consider each component separately, to understand its function fully and to make improvements through better arrangements and materials'. As the historian Michael Bailey explains, Stephenson sought to improve boiler design and construction, eliminate the loss of heat from boiler to cylinder, and to enhance both transmission and suspension.[5]

Through this work, Robert Stephenson was the architect of a new geometry for railway locomotives, which found expression in his Planet, completed in Newcastle in early September 1830, less than two weeks before the opening of the Liverpool to Manchester line. 'Stephenson's programme to improve the locomotive from the mineral railway "travelling engine" to the faster, more economic and more powerful Planet type [was] achieved in just 13 months [, …] a remarkable achievement that ranks among the foremost of technological advancements', writes Michael Bailey.[6] Thereafter came the 2-2-2 Patentee, first supplied to the Liverpool and Manchester in 1834. But Stephenson freely admitted that his own Newcastle factory was not up to the task of making the components for these engines to the right standards of accuracy or scale.

---

4 Hills, op. cit., 191.
5 Michael Bailey, *Robert Stephenson, Eminent Engineer* (2003), 168.
6 Ibid., 171.

Nor were others: in 1834, the reputable London firm of G. & J. Rennie was commissioned to build five engines, but they were so poorly constructed that they had to be rebuilt after a year by the Manchester firm of Fairbairns, where the machine shop was largely tooled by Whitworth.[7] The Maudslay men had been indirect beneficiaries of the railway boom already, participating in the surging demand for all engineering products, but now they had the opportunity to supply the industry directly, and in doing to, to shape its future course.

By 1840 there were six locomotive manufacturers in Manchester alone, with others in London, Liverpool, Crewe and Newcastle. In that year, an estimated 279 engines were built. But as the railway boom gathered momentum, it depended on and reinforced the innovations introduced by the Maudslay men. Standardization, accuracy and mass production all came together to meet the needs of the railway system. Both Roberts and Nasmyth became significant manufacturers of locomotives in their own right and, together with Whitworth, they supplied machine tools and machined components (wheels, axles, fireboxes, boiler tubes and the like) to the industry. The firms became in themselves nurseries of talent, replicating the success of Maudslay's original factory.

* * *

In 1838, Richard Roberts told a journalist that his attention for some years past had been directed to 'the practical workings of the railways,' particularly the Liverpool and Manchester line. His firm had been first, or at least joint first, to build a locomotive in Manchester, building the Experiment in 1833 (the rival claim comes from William Glasgow's Caledonian foundry). Roberts patented aspects of the new design, and built a second experimental engine along similar lines, the Hibernia, for the Dublin and Kingstown railway. Neither engine was a great success, and for some years he was preoccupied with building a steam carriage for use on the roads. Like Trevithick and others before him, the task defeated him: in April 1834 his model was given a trial run on Oxford Road, where it got up to 20 miles an hour before the boiler exploded, covering its 50 passengers with soot, and its inventor with ignominy. In 1837, having employed the talented German engineer, Charles Beyer, he resumed making steam locomotives. The following year, he and his backers decided to build a new factory dedicated to railway engineering.

---

7 See Norman Atkinson, *Sir Joseph Whitworth: The World's Best Mechanician* (1996), 38.

This became the Atlas Works, built on Oxford Road adjacent to the site of the original Globe Works on Faulkner Street. Roberts designed a large clock that was installed over the new factory, a landmark for generations of Mancunians and a boon to a workforce obliged to live their lives subject to precise timekeeping. 'Time is money,' was no empty phrase: at this stage of the Industrial Revolution, time worked was monitored with as much attention to detail as other factors of production.

'It was [...] especially in the quality of their workmanship that the engines of 1830–37 needed improvement,' noted a contemporary observer, 'and to this improvement the practice of Mr Roberts greatly contributed....] He introduced superior workmanship, giving better proportions and finish to the principal parts[.... He] delighted in refined workmanship in connection with every kind of mechanism and he was the first who brought to bear on the modern locomotive that kind of craft which, while it did not alter its general mode of action, nevertheless contributed so much to its working economy. His framing was stronger and the fastenings better. His bearing surfaces were larger and better got up; and in all parts machine work was substituted to a great extent, in place of hand fitting more commonly used by the north country engineers[. ...] Mr Roberts added characteristic features of design, which still distinguish the English locomotive engine; and by his practice, he sensibly advanced the standard of locomotive construction upon every railway in the kingdom'.[8] Samuel Smiles was the first to point out how Roberts achieved this. In essence, he deployed the know-how acquired in the mass-production of textiles machines. 'His system of templets [sic] and gauges, by which every part of an engine or tender corresponded with that of very other engine or tender of the same class, was as great an improvement as Maudslay's system of uniformity of parts in other descriptions of machinery.' These templates and gauges enabled the replication of precise measurements and thus facilitated the interchangeability of components. His unsung advances in this respect prefigure the industry-wide standardisation of nuts and bolts and measures introduced by Joseph Whitworth.

Richard Roberts was associated with the railway business until June 1843, following the death of his partner Thomas Sharp in the previous year; over a period of a decade, the firm received around 80 orders and manufactured nearly 250 locomotives, of which 111 engines were shipped to the European continent. At this point, the firm 'demerged', Roberts retaining his interest

---

8 Cited in Hills, op. cit., 194.

in the Globe general engineering works, while the remaining partners took on the railway business and developed it into a globally successful company, manufacturing 1,500 locomotives by the time of Roberts's death in 1864. Charles Beyer perpetuated the Maudslay–Roberts line of manufacturing excellence when he left the firm in 1853 to go into partnership with Richard Peacock, a locomotive engineer, thus establishing Beyer Peacock, one of the world's great engineering companies, which carried on in business until 1966. John Ramsbottom (1814–97) of Sharp Roberts went on to become chief engineer of the Manchester and Birmingham Railway and subsequently the London and North Western Railway (LNWR), created in 1846 with the amalgamation of the Grand Junction, the London and Birmingham and the Manchester and Birmingham railway companies. Ramsbottom became a hugely influential figure in British engineering, presiding over the LNWR's vast locomotive works at Crewe. The Atlas factory set the standard for steam locomotive engineering for a century.

Roberts himself no longer had a direct interest in railway business, but in 1846 he was called upon to solve a major manufacturing headache encountered during the construction of Robert Stephenson's tubular bridges over the River Conway and the Menai Straits. Stephenson, who was chief engineer to the Chester to Holyhead railway, conceived of the idea that the bridges should be made out of hollow wrought iron tubes, inside of which the locomotives would eventually run, but he was too busy overseeing other projects, appearing at parliamentary hearings and prospecting for new business overseas, to take day-to-day interest in their design, delegating the practical work to the Manchester industrialist William Fairbairn. By this time, Fairbairn had a shipyard in Millwall, on the north shore of the Thames on the Isle of Dogs, as well as extensive business interests in Manchester, and it was in London that he built and tested the wrought iron tubes to ensure that they would be safe and strong enough to carry the heavy load. From building iron ships and boilers, Fairbairn knew that to achieve the optimal strength, the tubes would have the same fundamental characteristics of a hollow girder, albeit on a massive scale, each section of the bridge weighing in at 1,550 tons. As so often with the great nineteenth century engineering projects, it was necessary to embark on a series of experiments to establish the optimal structure in practice. Fairbairn bashed and smashed his model girders to destruction, and there ensued a public dispute between him and his erstwhile friend, the mathematician Eaton Hodgkinson, who was brought on board to provide a theoretical undergirding to Fairbairn's empirical approach. Stephenson left Fairbairn and Hodgkinson to slug it out, '[hating] each other most enthusiastically', as one contemporary put it. Both men claimed credit for a crucial aspect of the final design, a layer of iron cells on the top of the tubes that prevented them from puckering or buckling. Eventually, Fairbairn resigned, publishing a detailed and self-justifying account of his experiments.[9] This was after the first trains had rolled across the Conway Bridge, and before the same technology was used for the bigger and bolder Britannia Bridge over the Menai Straits nearby.

Having established the perfect structure, albeit without understanding quite how it worked, the challenge was to physically construct the tubes, and to do so under enormous time and cost pressure. This is where Richard Roberts came in, solving the practical manufacturing problem. The tubes were made up of plates of iron 12 feet long by 2 feet 8 inches wide and were joined together using rivets. Before Roberts, a single hole required several men to mark the hole and line up the plate, and the result was invariably inaccurate. The holes in one iron plate never seemed to line up with the ones on the adjoining plate. Workmen would get out the file in the hope of making a fit, with the result that the hole would end up being either much too big for the rivet or could only admit the rivet if inserted at an angle. The rivets had to be introduced when red hot, and the result of this bodging was that the final structure would be weak and poorly finished. Humans simply could not do this boring and repetitive – not to mention strenuous – work, and Roberts was called on to develop a solution.

Within a matter of weeks or, according to some stories, in the course of just one evening, Roberts designed his 'Jacquard' punching machine. This was the first digital machine tool, deploying the ideas pioneered in the Jacquard loom, an ingenious French invention that allowed patterns to be woven automatically. For the loom, patterns were made on pieces of card that, like a roll of music inserted into automatic piano, dictated the way the machine operated. Insert a different card, and it would weave another pattern into the cloth. Babbage seized on this in his design for his Analytical Engine, the successor to his Difference Engine, and Roberts did the same for this pioneering machine tool, which allowed the rivet holes in a sheet of iron to be punched 'to any required pattern and pitch without need of templates or marking, and with a precision enabling any two of a thousand plates to be riveted together without the subsequent correction of the rivet holes, which had always before been necessary.'[10] The machine, which Roberts patented in March 1847, could

---

9 William Fairbairn, *An Account of the Construction of the Britannia and Conway Tubular Bridges* (1849).
10 *Illustrated London News*, 11 June 1864.

punch 22 plates or 3,168 holes in one hour, to such a standard of accuracy that
if the plates were stacked on top of one another a solid iron bar could be passed
all the way through the holes, and any two plates could be riveted together.
Three men and a boy operated the machine. 'Your Multifarious Punching
Machine was used at the Conway Tubular Bridge,' wrote a clearly delighted
Robert Stephenson in a personal letter to the inventor. '[I]ts advantages over
the machines in common use, both as regards accuracy, speed and economy
were perfectly established.'[11] The first tube was floated into place in March
1848 and opened to traffic on 1 May.

Queen Victoria visited the Britannia Bridge, which was constructed on the
same principles, in October 1852. 'From the Menai Bridge, itself a wonder,'
she wrote, 'the stupendous Tubular Bridge is seen to great advantage. A little
way on, we got into the train, being met by the celebrated Mr Stephenson
himself. We went quite slowly till we got to the bridge itself, where we got
out, going to its entrance to examine its wonderful construction. Two gigantic
ancient Egyptian lions stand on either side of the entrance. The whole, is
of Egyptian architecture, – extremely massive and handsome. The 2 tubes
are unconnected, which as Mr Stephenson explained to us, is much safer.
The tube is 180 yards in length. Albert walked over the top of it, while the
Children & I were pulled through by hand in a railway carriage. We got out
again and walked down to the water's edge to see the construction from below.
The height is immense, 105ft:, the middle tower being as high as the London
monument! The Menai Straits are at this point, full of dangerous currents and
eddies, the tide running 7 miles an hour.' After listening to a lecture on the
difficulties of erecting the bridge under such conditions, the monarch returned
home to Windsor, en route passing through Wolverhampton, which in contrast to Manchester or Liverpool, she declared to be 'one of the most dreadful
parts of the country one can imagine.' The desolate iron workings, the piles
of slag heaped like mole hills, the dilapidated cottages and the thick and black
atmosphere lit up by the flames of the forges, the landscape devoid of all but
a few leafless trees, chimneys as far as the eye could see – this seemed another
world. 'In the midst of so much wealth, there seems to be nothing but ruin,'
she reflected.[12]

Roberts's technology was used in the construction of the Boyne Bridge in
Ireland, the Victoria Bridge over the St Lawrence at Montreal (which deployed
tubes of over a mile and a quarter long) and the Jamuna Bridge of the East Indian

11 Richard Hills, *The Life and Inventions of Richard Roberts* (2002), 106.
12 Queen Victoria's Diary, 14 October 1852.

Railway. Once again, despite his revolutionary invention, Roberts derived no long-term pecuniary benefit. His involvement with railway technology could be playful, as well as commercial: he built a 'loop-the-loop' railway to illustrate centrifugal force for the delight of his children: 'It was in the form of a railway with a little carriage which could be filled with water; this, running down an incline, gathered speed, turned upside down without spilling any water and ran up an incline on the other side.' One of the toys was preserved by the Manchester Literary and Philosophical Society but was destroyed in the Blitz.

* * *

James Nasmyth had long been interested in steam transportation: he had built a steam vehicle while still in Edinburgh and had attended the Rainhill trials (carefully making a drawing of the locomotive he mistakenly thought was the Rocket). His Bridgewater Foundry entered the locomotive business at the same time as Roberts. In the 21 years from 1836 to 1856, during which Nasmyth was managing partner, the firm built 109 steam locomotives, almost all sold to British companies. This accounted for sales of some £231,000, a quarter of the firm's total, and further evidence of Nasmyth's commercial versatility. The extra string to the firm's bow helped it withstand periodic downturns in other markets, and vice versa – for example in the late 1840s, when the railway market crashed and companies that were not broadly diversified went out of business. Nasmyth sensed early that the market would be enormous and took the considerable risk of building four locomotives 'on spec' in order to acquire the necessary specialist expertise and demonstrate the firm's capabilities to potential buyers. In August 1840, the firm received a massive order for 20 engines and tenders from Brunel's Great Western Railway; of these, 16 were passenger engines delivered between July 1841 and December 1842. Daniel Gooch used the Actaeon for the opening of the Bristol and Exeter line on 1 May 1844, and himself drove the train from Exeter to London and back. Nasmyth devised special machine tools for his railway customers: 'these tools [...] rendered us more independent of mere manual strength and dexterity, while at the same time they increased the accuracy and perfection of the work. They afterwards assisted us in the means of perfecting the production of other classes of work. At the same time they had the important effect of diminishing the cost of production.'[13] Nasmyth claimed in his *Autobiography* that Gooch

---

13 Nasmyth, *Autobiography*, 229.

provided an unprompted testimonial letter as to quality of the work, but that is an overstatement: Nasmyth in fact wrote an unctuous letter asking 'might I beg you as a most valued favor that you would favor us with a few lines of your performance of the engines[. . . .] I need not say how truly valuable and important to us would be a few words coming from so high an authority as yourself'.[14]

With the railway mania of the mid 1840s, the market for locomotives expanded dramatically, and Nasmyth won a substantial order from Robert Stephenson, shipping 27 engines and 25 tenders to Stephenson's Newcastle works between October 1845 and November 1847. There were big orders from the LNWR and the Great Northern Railway as well. By the late 1840s, locomotives accounted for 40 per cent of the firm's business. In truth, the company was one of the United Kingdom's smaller locomotive manufacturers, producing on average of just seven engines a year between 1838 and 1853. But there were also burgeoning sales of steam hammers: Nasmyth manoeuvred himself into the enviable commercial position in which he could sell to all parties in the railway industry, including competitors such as Sharp Stewart, Beyer Peacock and Stephenson's factory in Newcastle, as well as rail companies such as the LNWR, which ordered 14 hammers from Nasmyth before he retired.

By the end of 1845, 623 railway schemes had been laid before Parliament, with a projected capital outlay of £563 million, a gigantic amount of money. Not all the schemes were built by the time the boom came to an end a mere three years later, but 6,084 miles of track were in operation by 1850, and by 1880 'total investment in that industry came to more than half the country's GDP, comparable to $8 trillion for the US today,' notes the economic historian Andrew Odlyzko.[15] Throughout Europe, the 1840s was a time of economic depression, social unrest and political upheaval, culminating for many countries in fully fledged revolution and subsequent repression. King Louis-Philippe of the French was deposed in February 1848, amid bloody riots, and later that month, Prince Metternich of Austria was overthrown. These were the 'hungry forties', and Great Britain was hardly immune to the turbulence affecting the Continent. Karl Marx and his Manchester mill-owning patron, Friedrich Engels, published their *Communist Manifesto* in the same month, hoping to stir

---

14 Cited in Cantrell, op. cit., 191–92.
15 Andrew Odlyzko, 'Crushing National Debts, Economic Revolutions, and Extraordinary Popular Delusions', University of Minnesota Working Paper, 5, to be found on http://www.dtc.umn.edu/~odlyzko/doc/mania05.pdf

up revolution in their adopted homeland, and indeed the Chartist movement of working-class agitation culminated in a petition signed by more than one million, calling for universal suffrage and other reforms, and delivered to Parliament on 10 April. Tens of thousands of disaffected workers took the train, that most new-fangled and disturbingly democratic form of transport, to the metropolis, gathering at various locations with a plan to march on the House of Commons. The Whig government panicked, and recalled the Duke of Wellington from retirement to forestall the threat of violent disorder. The duke ordered eight thousand troops to the capital, where, together with tens of thousands of volunteer special constables, they were successful in neutralizing the threat of revolution. They sealed off the bridges, permitting the proletarian protesters to cross the Thames only at Blackfriars. The vast body of men was headed off to Kennington Green in the South, and the threat of insurrection was averted.

The huge boom in railway investment was a countervailing force, arguably the catalyst to the next stage of the Industrial Revolution, when Britain's productivity and prosperity accelerated and its global economic hegemony was unchallenged. Odlyzko notes that 'in the peak year of 1847, direct employment just in construction of new lines involved an army of manual workers that was more than twice as large as the British army[....] On top of that were the spillover effects from goods and services provided to these workers[....] It was a pseudo-Keynesian stimulus [that produced] a supply-side shock to the economy that compensated for the negative effects of famine and of disruption in foreign trade'. It defanged the social tensions that were tearing apart other European countries.[16] Benjamin Disraeli wrote that prosperity helped 'make the Chartists forget the Charter'.

The impact of the railways was more than purely economic: with the new speed of travel, contemporaries felt that the traditional constraints of time and space had been annihilated, and indeed the ancient variations in time across the country were eliminated with the introduction of standard railway time: hitherto the time changed by 4 minutes with every degree of longitude, meaning that Birmingham was 7.25 minutes behind London, and Penzance and Walton-on-the-Naze more than half an hour apart. (Greenwich Mean Time became the legal standard for the United Kingdom only in 1880, but the exigencies of the railway timetable meant this had been achieved in practice decades before.) In the age of the stagecoach, the average speed for a long journey was no more than 10 mph, compared to more than 50 mph for

---

16 Odlyzko, op. cit., 6.

trains. The *Sunday Times* observed that it was at last possible for an individual to have breakfast in London, 'dine in Birmingham, transact business, and sup in the metropolis during the course of the same day'.[17] The journeys taken by the working men of the North to London to deliver the Chartist petition, or by the farm labourers of Hardy's Wessex to visit the Great Exhibition three years later, were unprecedented. As Thomas Hardy wrote, the coming of the railways brought about more social change for Great Britain than any event since the Norman Conquest.

* * *

If the start of the railway boom coincided almost exactly with the accession of the new Queen, the same was true of the advent of the transatlantic shipping industry. The Maudslay-powered SS *Great Western*, the 'Queen of the Atlantic', and for two years the largest passenger ship in the world, was launched at the Floating Harbour in Bristol on 19 July 1837, less than a month after Victoria became Queen (Figure 6). The ship was dreamt up at a board meeting of the Great Western Railway at Blackfriars in London, in October 1835, when Isambard Kingdom Brunel responded to a complaint that his London to Bristol railway line was too long, by asking out loud: 'Why not make it longer? Build a steamship to go to New York, and call it the Great Western?' Taken as a joke by most of the railway promoters in the room, the sugar entrepreneur Thomas Guppy determined to support the young Brunel, and the project became a reality.

The *Great Western*'s hull was entirely traditional, built of oak by the firm Patterson & Mercer of Bristol, but the conception was new. When she left her home port on her maiden voyage, she was a pure sailing ship with a tonnage of 1,200 and 212 feet in length, 35 feet across but broadening to just under 60 feet if the paddle boxes were included, with four masts that could be rigged with main, mizzen and topsail like a schooner. It was intended that her yet-to-be-installed Maudslay side-lever engines would be the main source of power, to be used continuously when at sea, while the sails would be rigged only when the winds could deliver extra speed, or to provide balance in choppy seas and keep the vessel in trim: a considerable psychological and technological step forward, as hitherto paddle-steamers were built so that the engines provided auxiliary power only, and could be disengaged when winds were favourable. Thousands gathered to watch the launch of the ship with the largest keel in the world. 'At five minutes past ten, the dog shores having been struck away,

---

17 *Sunday Times*, 23 September 1838.

FIGURE 6 SS *Great Western*, Brunel's first transatlantic vessel, powered by Maudslay engines (Courtesy of the British Postal Museum and Archive, London)

the screw was applied, and a general shout arose, – "she moves,"' reported the *Bristol Mirror*,[18] 'which indeed she did, in the most majestic and graceful manner. For the moment all was hushed, whilst the magnificent and beautiful vessel glided into the water[....] At the calculated distance she was checked by a chain cable, and brought up within a few feet of the opposite shore, without the smallest accident. The launch being concluded, loud, spontaneous and continued cheers took place, and the immense crowd began to disperse.'

With a little help from tugboats, the *Great Western* made her way out of the River Avon to the English Channel and all the way along the south coast into the Thames, her first destination being the East India Docks of London, where she would be equipped with steam engines. Maudslay, Sons and Field had been appointed on the recommendation of Brunel because of the firm's record as the leading manufacturer of large marine engines. 'I think you will be safest,' wrote Brunel to his board, 'in the hands of the parties who have the most experience, and [...] Messrs Maudslay are those persons. Their price is, I think, moderate'.[19]

18　*Bristol Mirror*, 22 July 1837.
19　Cited in Denis Griffiths, *Brunel's Great Western* (1985), 16–17.

The pistons, cylinders, cross-heads, four huge boilers, rods, crank-shafts, steam pipes and water-cocks, pumps and other components for the engines were shipped by horse-drawn cart from Lambeth to Blackwall, where they were hoisted on board by derricks and assembled in the 80-foot-long engine room by an army of mechanics. The crankshafts linking the engines to the paddles had been made by Acramans of Bristol and were brought to London as cargo to be assembled. An old hulk was berthed alongside to serve as a floating workshop. The *Great Western* would be fitted with luxurious cabins and a passenger saloon that ran the full length of the vessel. She would have the capacity to accommodate 128 passengers and 60 crew.

Many doubted that the voyage could be made at all. Dionysius Lardner, the popularizer of the great technological advances of the day, was to make a fool of himself for all time when, in late 1835, he stated that 'as to the project of making a voyage directly from New York to Liverpool, it was, he had no hesitation in saying, perfectly chimerical, and they might as well talk of making a voyage from New York or Liverpool to the moon.'[20] Lardner was not as idiotic as this statement suggests, though he did predict that trains would not be able to run at speed because of the risk of asphyxiating the passengers, and he was equally wrong in his belief that the doubling of a ship's size required a doubling of its engine power. Using elaborate but erroneous calculations, he claimed that the maximum range for a steam ship was 2,550 miles and therefore that a ship could not cross the Atlantic under its own steam (quite literally), but would have to make a stop on the Azores, or try a shorter route, as for example, from Valentia Island off the west coast of Ireland to Newfoundland. But as Brunel was about to prove, and as other shipbuilders such as Macgregor Laird, one of the Laird brothers who built the docks and town of Birkenhead, also knew from experience, bigger ships were more efficient. 'It is well known that the proportionate consumption of fuel decreases as the dimensions and power of the engines are increased,' Brunel explained, 'and consequently that a large engine can be worked more economically than a small one.'[21] Brunel and the Lairds understood that the amount a ship can carry increases in proportion to the cube of its dimensions, while water resistance increases as the square of its dimensions, meaning that a bigger vessel can travel more quickly, or the amount of power can be reduced.

After successful trials, the *Great Western* made its inaugural voyage on 31 March 1838, leaving Blackwall at just after 6 p.m. Some sixty passengers had

---

20 *Liverpool Albion*, 14 December 1835.
21 Griffiths, op. cit., 15.

booked for New York and were to be picked up at Bristol. On board for the first leg of the 670-mile passage to Bristol were the 68-year old Sir Marc Brunel and his son Isambard, as well as Joshua Field and Joseph Maudslay, who were to superintend the working of the engines. All went well until just past Gravesend, by which time Sir Marc had disembarked, when crew members noticed the smell of burning oil emanating from somewhere near the base of the funnel, quickly followed by flames and dense smoke. The captain beached the ship on mudflats, and a group of stokers abandoned ship, telling everyone they met onshore that the *Great Western* was doomed. Meanwhile, George Pearne, the chief engineer, made his way down into the smoke-filled engine room and managed to take some pressure off the boilers, forcing water into them and thereby averting an explosion, while Isambard Kingdom Brunel and others got a fire engine working on deck. Brunel, it will be remembered, nearly lost his life under the River Thames during the construction of the tunnel more than a decade before; now, he suffered another serious accident. Pearne described what happened after two engineers had descended to the engine room:

> We got the engines and hand pumps to work and all hands baling, pumping, etc, succeeded in extinguishing the fire. The most melancholy part of the catastrophe was that I.K. Brunel, in attempting to go down the fore stoke-hole ladder, stepped on a burnt rung, several of which, in this state, giving way, precipitated him down to the bottom about twenty feet, falling on Mr. Claxton [of the Great Western Steamship Company.] He was taken up, apparently seriously injured, and ultimately sent ashore.[22]

Brunel was knocked unconscious, having fallen facedown into a pool of water, and Claxton undoubtedly saved his life, by breaking his fall and pulling him out of the water. Despite appearances, however, neither the ship, nor its designer, suffered long-term harm. Brunel was put ashore at Canvey Island in the Thames Estuary, and, after a few days in bed, he was back on his feet. The ship was floated off on the next tide, its engines in full working order. She steamed towards Bristol under the watchful eyes of Maudslay and Field, arriving at the mouth of the Avon at 6 a.m. on 2 April. The trip had taken 58.5 hours, including 6.5 hours delayed by the fire, an average speed of an outstanding 13 knots.

---

22 Christopher Claxton, *The Logs of the First Voyage [...] between England and America, by the Great Western* (1838), 1–2.

Over the next five days, some 3,000 people came to inspect the ship at its mooring seven miles downstream from Bristol, many of them amazed to see that she had made the voyage at all. When she finally did set sail for the United States on 8 April, a day later than advertised, she was carrying only seven passengers, the rest having cancelled because of the fire. But they need not have worried: the ship completed the passage in 15 days, with 203 tons of coal unburned. It was not quite the first steamer to make the voyage, as the much smaller, Laird-backed *Sirius* arrived in New York the day before Brunel's ship, carrying 97 passengers. Yet (according to propaganda circulated by those close to Brunel, at least), the *Sirius* had been compelled to burn furniture, a mast and yardarms to make the passage. Certainly, the *Great Western* proved herself faster and more efficient, making a further 44 crossings in its eight years of service. A new era of international travel opened up, driven, quite literally, by Maudslay's giant side-lever engines. Amid the publicity that surrounded Brunel's achievements, Maudslay's contribution was often overlooked, but for engineering aficionados the engines were deemed the *ne plus ultra* of their type. 'Machinery of this quality is able to speak for itself,' noted the *Artisan* in 1844, 'we do not know of any production in steam machinery that is in every way more creditable.'[23]

---

23 Quoted in Griffiths, op. cit., 77.

CHAPTER 10

THE TURN OF THE SCREWS – SIR JOSEPH WHITWORTH AND THE QUEST FOR MECHANICAL PERFECTION

Thomas Carlyle had profound views on the impact of industry on Victorian society, but he rarely met an engineer in the flesh. Whenever Carlyle and his sister, Jane, encountered Joseph Whitworth, Nasmyth's great rival, they sniggered at his broad Lancashire accent, scruffy appearance and common manner. 'He looks cousin once removed from a baboon,' wrote Jane Carlyle in her diary after Whitworth dropped in for tea at the sage's Chelsea home. 'He is still the *mechanic* in appearance and bearing; all the perfumes of – the Bank of England cannot wash *that* out of him.'[1] In other words, he was still a plain working man, despite his immense wealth. Another time, Carlyle himself was impressed by Whitworth's 'sloping brows, prominent eyes [...] head like a dog's rather'.[2]

Among the stiff portraits of frock-coated and bewhiskered Victorian industrialists, Whitworth does look peculiar, with high cheekbones, big eyes, and sprouting white hair. He also looks disappointed, possibly a reflection of his troubled childhood or his frustrating interactions with government, or more innocently a result of the freezing effect of the long exposure times required by Victorian photography. For all the mockery, the sage and his sister were deeply impressed at his energy and practical intelligence. 'When one talks with him one feels to be talking to a *real live man*.' The philosopher and his sister

---

1 *Jane Welsh Carlyle's Journal*, 18 November 1855. See the Carlyle Letters Online: http://carlyleletters.dukejournals.org/cgi/content/full/30/1/ed-30-jane-welsh-carlyle-journal accessed on 27 June 2015
2 Letter from Carlyle to his brother, John, 19 November 1855

were especially taken by Whitworth's invention of the mechanized besom, or automatic street-sweeping machine, which was rapidly adopted in the 1840s and helped clean up Victorian England's filthy cities. In the era of frequent cholera epidemics and profound urban squalor, there was from the 1830s onwards a high-minded debate about sanitary engineering. Edwin Chadwick, the most prominent of these campaigners, was a partner in the enterprise to sell Whitworth's mechanized besom. These devices look like World War I tanks and were in fact horse-drawn: a series of brooms some 2.5 feet across were attached to a pair of endless chains suspended from a wrought iron frame. The dirt and detritus were swept up into a cart, thus doing the degrading work typically carried out by hand by the lowest of the low of Victorian society (e.g., Little Jo in Dickens's *Bleak House*). With his typical eye for detail, Whitworth calculated that the machines cost a mere 10–15 shillings per week to operate, about a third of the cost of the human equivalent.

Notwithstanding the impact of the street-sweeper, Whitworth was not celebrated for one startling invention, but rather for 'the general stamp of excellence, which he has been enabled to impress on the machinery of the United Kingdom'. As his contemporary, Sir James Emerson Tennent observed: '[W]hat he found rude and complete, he rendered as nearly as possible perfect, till there is scarcely an operation connected with the shaping of metal, in cutting, planing, turning, boring, drilling or slotting it, to which he has not applied machines in supersession of hand labour, such as the world never saw before; unsurpassed in excellency, in design, and in perfection of execution.'[3] Nasmyth, Roberts and others designed, built and sold machine tools, but Whitworth dominated this field, constructing machines that were sturdy, heavy and cleanly finished, celebrated for their precision and reliability. These were marketed by means of catalogues of photographs, which can be seen at the Institution of Mechanical Engineers: the turning lathes for railway engines, the locomotive drilling machine, the universal horizontal boring machines and punching and shearing machines are presented with the sublime self-confidence of the undisputed market leader. 'All prices are subject to alteration without notice,' the catalogue states, implying that customers would be lucky to take possession of such machines on any terms whatsoever. Each mechanized drill or lathe or slotter sold continued the mechanical reproductive cycle first set off in Maudslay's and Clement's works in South London. At Maudslay's, he perfected the true plane and imbued the

---

3 Sir James Emerson Tennent, *The Story of the Guns* (1864), 23.

philosophy of mechanical perfectionism, while at Clement's his work on the Difference Engine taught how to make identical items in large volumes. 'The very soul of manufacture is repetition,' he urged.

Whitworth had arrived back in Manchester in late 1832, opening a workshop in Port Street, hanging a sign on his door: 'Joseph Whitworth, tool maker from London'. He did not have Nasmyth's connections with investors and customers, and in the absence of records, one suspects that his business grew like that of his friend, William Fairbairn, who describes in his autobiography how he worked night and day, seven days a week, through the 1820s and 1830s, his products in constant demand, financing the growth with retained profits and modest bank loans. In 1833, one year after Whitworth moved back to Manchester from London, he moved to five-storey premises at 44 Chorlton Street, on the banks of the Rochdale canal. The business started modestly, with outgoings of £50 a month implying a workforce in low double figures. Between 1834 and 1839, he took out 15 patents for machine tools, including: the hollow box structure, the duplex lathe, and 'the quick return motion and reversing tool in planing machines', as well as improvements to many others.

Growth was rapid, and Whitworth employed nearly 400 people after a decade, 636 by 1851, and more than 1,500 later in the century, by which time he had opened a second factory in Openshaw on the outskirts of Manchester. The volume of machinery produced rose from 50 tons a week in 1842 to 200 tons by the time of the Great Exhibition, and more thereafter, as he enjoyed a boom in orders. The factory itself became celebrated: Prince Louis-Napoleon (later Emperor Napoleon III) visited in January 1839, and Prince Albert came on his trip to Manchester in late 1851, as did Prime Minister Kossuth of Hungary. Whitworth threw open the works to all who were interested, maintaining a showroom full of model machines.

The factory was pioneering in many ways, meticulously laid out, with novelties like a mechanized gantry crane and a hydraulic lift. It was clean and orderly, paved with square flagstones of identical size so new machines could most easily be installed. 'It [also had] an arched semi-circular roof of a rather novel construction,' noted the *Engineer*. 'The beams have been so constructed as, without the aid of tie-rods or struts. To throw the weight perpendicularly on the side-walls, thus affording a clear and unobstructed span of about 40 feet under which a travelling crane passes, supported on ledges a short distance below the junction of the roof and the side-walls.'[4] The Crystal Palace used

---

4 The *Engineer*, 6 June 1856 and 14 December 1866.

similar techniques and had a 72-foot span, as did eventually the Royal Albert Hall, with a 219-foot span. Charles Beyer, at the time working for Richard Roberts and subsequently a great railway engineer in his own right, visited in July 1842 and was deeply impressed:

> Those works appeared to me [...] a genuine pattern for any kind of machine-making establishment to arrange and conduct them by; our own (Sharp Roberts) by no means excepted. The pains taken in doing work is as great as their means they employ as well as ways upon which they seem to proceed is judicious. The fitting and filing work generally is superior to anything I ever saw before. Their principal tools are lathes, planing machines and all sorts of screwing tackle generally; drills; and now they are commencing to make key-groove machines. Having made several additions to their works since the erection of the main building they have laid down in the yard a turntable from which branch rails (radiate) for the better connecting of the different premises.[5]

Whitworth had accumulated an unrivalled concentration of mechanical power, with more than a dozen planing machines carrying out the most massive work.

In these years, in contrast to the decades of fame after 1851, Whitworth kept a low profile. He joined the appropriate Manchester societies, for example, the Lit and Phil, and in 1841 was admitted to the august Institution of Civil Engineers (of which he was president in 1853). He busied himself with his favourite social causes, for example technical education for the working classes. He acquired a substantial home with 52 acres of grounds at The Firs, in Fallowfield, where he entertained fellow industrialists and the politicians Richard Cobden and John Bright, who led the successful campaign to abolish the Corn Laws. But from the 1840s, when the success of his business was not in doubt, and his own financial position secure, he began to take a higher national profile. A rich and influential man who was not afraid to wade into controversy, Whitworth became a prophet of precision, giving lectures on his pet subject in museums and mechanics' institutes up and down the country, bringing with him caseloads of machinery that he would unpack and demonstrate in action with the panache of a showman. Dour, practical, preoccupied with detail, Whitworth was an unlikely master of propaganda.

---

5 Atkinson, ibid. 127.

He promoted what became known as the Whitworth system of manufacturing, the elements of which were: the true plane, measurement by touch, standard measures, decimalisation and interchangeable components, particularly nuts and bolts and screws.

True planes did not look especially impressive when seen in situ on a workman's bench, as even Whitworth conceded, but they were 'the foundation of his entire system, the elements of mechanical truth [...] from them is fashioned every true machine.'[6] 'I cannot impress too strongly [...] upon all in any way connected with mechanism, the vast importance of possessing a true plane, as a standard of reference,' Whitworth insisted. 'All excellence in workmanship depends on it.'[7] The three identical surfaces used to establish a true plane were a sort of Holy Trinity or 'mystic triad' of mechanical accuracy, each perfect surface capable of begetting another and another, ad infinitum.

You can only make as well as you can measure, he used to say. To that end, Whitworth told the Institution of Mechanical Engineers that engineers should think in tenths, hundredths and thousandths of an inch, rather than eights, sixteenths and thirty-seconds. 'If we had a better system of notation for our measures,' he urged, 'the importance of minute and accurate measurements would become more familiar.'

The machines in Whitworth's factory were calibrated to the ten-thousandth part of an inch. Measurement had to be made, not by a ruler or caliper, but end-to-end between two perfect planes, using touch rather than a rule to attain levels of precision that Maudslay could have only dreamed of.[8] 'If an object be placed between two parallel true planes,' he explained, 'adjusted so that the hand can just feel them in contact, you will find on moving the planes only the fifty-thousandth of an inch closer together, that the object becomes distinctly tighter, and requires greater force to move it between them.' This millionth measuring machine was a micrometer modelled on Maudslay's Lord Chancellor. Smaller than a sewing machine, it attained new and unimaginable levels of precision (an original is preserved in the Science Museum). Whitworth would bring out the device to display to his wondering audiences how a loosening of one millionth of an inch was enough to cause an object to drop to the floor, gravity overwhelming friction. Conversely, he showed how when he touched the bar with one finger, the temperature would rise by one seventh of a degree Fahrenheit, causing the bar to expand by 1/100,000 of an inch.

---

6 The *Engineer*, 6 June 1856.

7 Lecture to the Institution of Mechanical Engineers, Glasgow, 1856.

8 Cited in Tennent, op. cit., 26.

Nasmyth joked privately, and not without a hint of professional jealousy, that the machine was so delicate, that its readings went askew the moment more than one person looked at it at the same time. Whitworth kept the machine in a glass case, with an opening big enough only to turn the micrometer wheel and insert the gravity piece.[9] But Nasmyth's sarcasm was not far off the mark: practical engineers knew that the new machine was too sensitive for everyday use. As propaganda, however, the millionth machine was unrivalled.

Decades before, Maudslay had had a similar idea but had not been able to achieve the necessary level of precision. He would have understood how, as Whitworth explained, 'it is of great importance to the manufacturer who makes parts of machines in large quantities to have the means of referring to an accurate fixed measure.' Through a painstaking process, Whitworth's machine was used to fashion standard yards, feet and inches, which could then be used to create standard gauges, the reference point for mass-produced components. These are so-called 'go' or 'no go' gauges which give a very practical demonstration of minute adjustments in measurement. For example, at one lecture Whitworth displayed three gauges: an internal cylindrical gauge with a diameter of 0.5770 of an inch, a matching external gauge of the same diameter and another a fraction smaller at 0.5769 of an inch. Though smaller only by one ten thousandth of an inch, this gauge was so loose and wobbly as to be no fit at all, while the two matching parts fitted perfectly. Vary the size in the other direction, and the part will simply not fit in. Until the advent of computerized machining, it was standard practice to use such gauges whenever a workman was boring a hole in a piece of machinery: '[T]he two gauges differing in size by a specific amount and the hole being only considered of satisfactory size when the one gauge will enter and the other not enter.'[10] The practical application was clear, for example, when making spokes for the wheels of steam trains or spindles for spinning equipment: instead of making each one by hand, one would arrive at a standard measure which would mean that components were interchangeable. The same principle was in time applied to making products such as engine components, locomotives, carriages, doors and window frames.

Whitworth's system was especially relevant when one manufacturer supplied parts of machines to be used by another. As Charles Babbage had found when trying to construct his Difference Engine in the 1820s and 1830s,

---

9 At a meeting of the Institution of Mechanical Engineers in Birmingham, on 27 July 1859, for example.

10 F. C. Lea, *Sir Joseph Whitworth: A Pioneer of Mechanical Engineering* (1946), 14.

parts had to be made in one place, otherwise they would not fit together. By the 1840s and 1850s, the demand for steamships and railway and mechanical engineering equipment was so enormous that individual factories could not cope, and cooperation between specialist engineers became essential. It was simply baroque to have thirty different ways of constructing a steam engine, or unlimited different types of nuts and bolts. The object was to have a system whereby 'every part of an engine or tender corresponded with that of every other engine or tender of the same class'.[11]

A key element was Whitworth's standardisation of screws. 'Great inconvenience is found to arise from the variety of threads adopted by different manufacturers,' the proselytizer of precision argued. 'The general provision for repairs is rendered at once expensive and imperfect [...] this evil would be completely obviated by uniformity of system, the thread becoming constant for a given diameter[....] As yet there is no recognized standard'.[12] He started by collecting samples of screws, nuts and bolts from many factories around the country and working out the average thread for different diameters. Based on these averages, he produced a series of screws at set sizes, all using the so-called 'Whitworth thread' with an angle of 55 degrees for the V-shaped thread and constant depth and radius. It was a diplomatic, as well as a practical achievement, to coax hundreds of competitors to adopt a common standard, testament to Whitworth's massive authority in Victorian England's ferociously competitive and individualistic business community. Each screw or nut and its corresponding bolt had a reference number and could be specified by its diameter and pitch, while its pitch would be checked using a hand-held gauge. He first proposed such a system as early as 1841, and the system was in universal use by the late 1850s. Railway companies, which had hitherto used a jumble of different threads, were early adopters. The British Empire at large followed suit, and the Whitworth Standard was officially adopted by the Board of Trade in 1880. 'The value of this single reform may be conjectured when it is borne in mind, that before his time every engineer and machine-maker provided his own screws, on no preconceived principle, but of the most arbitrary form and the utmost variety of dimensions. The consequence of this divergence was the utmost confusion, delay and expense in repairs.'[13] The screw could be replaced

---

11 Samuel Smiles, *Industrial Biography*, 271.
12 Joseph Whitworth's 'Paper on an Uniform System of Screw Threads, read at the Institution of Civil Engineers, in 1841,' published in *Misc Papers on Mechanical Subjects* (1858), 21–37.
13 Tennent, op. cit., 28.

with one conforming to the Whitworth standard. The system was in use in Britain and the Commonwealth until World War II.

There is a question mark about the absolute originality of Whitworth's achievements. Sometimes, as when in 1841 he lectured the Institution of Mechanical Engineers on a 'Uniform System of Screw Threads', he encountered sceptical harrumphs from Maudslay veterans: Joshua Field and Samuel Seaward both commented that there was nothing new in what he was saying, that it had all been invented long ago at the factory in Westminster Bridge Road. Maudslay, himself, had sought to develop a standard thread for screws, and others such as Roberts and Clement had continued their master's work. There was a degree of literal truth in the criticism from the old-timers: but their comments ignored the fact that Whitworth did more than anyone else to propagate Maudslay's ideals. Indeed, he did more than any other to *industrialise* the Maudslay system, to make the master's notions of precision engineering applicable across industry at large.

Whitworth also inherited Henry Maudslay's strong humanitarian concerns for the workforce, installing public baths near his factory, spending an increasing proportion of his fortune on technical education for workers. He celebrated the fact that mass production brought prices down dramatically: the cost of making a surface of cast iron true with hammer, chisel and file was 12s per square foot, compared to labour costs of less than one penny if a planing machine were used. Likewise, the price of a 29-yard bolt of printed cloth fell from 30s 6d to 3s 9d. This spectacular reduction in costs brought benefits to society at large, he contended. Staple goods became cheaper, and there would be more leisure time for workers, and less need for strenuous manual labour. The technology created new and better jobs for working people. This was a very different perspective from that of Nasmyth, for example, who saw his mission as eliminating troublesome humanity to the greatest extent possible from his own factory and those of his customers, while lowering costs and increasing profits for the entrepreneurial classes. Eventually, Nasmyth's detestation of the workers would drive him from industry to early retirement, while Whitworth would give away the bulk of his gigantic fortune to workers' causes. But for now, they and their peers were just too busy for sociological speculation. The great celebration of the machine age, and the acme of public recognition for the Maudslay men, came at the Great Exhibition of 1851.

CHAPTER 11

THE AMERICANS ARE COMING –
THE GREAT EXHIBITION AND
THE GREAT LOCK CONTROVERSY
OF 1851

Joseph Whitworth displayed no fewer than 23 machines at the Great Exhibition: the machine-tools, including lathes, planing, shaping, drilling and screw-cutting machines that were manufactured at Chorlton Street and shipped out by the ton, as well as the firearms that would increasingly come to preoccupy him. *The Times* noted that the Queen and Prince Albert made a 'minute observation' of Whitworth's millionth machine when they visited the Exhibition on Saturday, 7 June 1851, the newspaper vouchsafing that this machine 'illustrates in a higher degree, perhaps than any other object in the Exhibition, the extraordinary progress we have made in mechanism'.[1] Others who lingered in awe of it included Charles Dickens, not otherwise an enthusiast for the Great Exhibition. It was in the small side room housing the machine that Joseph Paxton, designer of the Crystal Palace (and archetypal self-made man, having risen from his humble position as gardener at the Duke of Devonshire's Chatsworth estate), introduced Dickens to John Bright, statesman and Rochdale millowner. Sadly, neither man made a note of the machine, Bright merely observing that 'Dickens's face scarcely indicates the possession of the powers of the mind he has displayed, tho' it is intelligent in expression, and pleasant to look upon'. Dickens recorded the antics of a party of schoolchildren rather than his impressions of the machine, while his novel *Bleak House* was intended as another kind of Great Exhibition, this one of social inequalities rather than mechanical expertise.

---

1 *The Times*, 9 June 1851. See also John Bright's diary entry for 24 May 1851.

Maudslay's won a prize for a marine engine, Nasmyth was rewarded for his steam-hammer, and Richard Roberts won prizes for two typically ingenious machines: a Patent Normal Drill (for drilling tiny holes in the frame plates of watches and clocks) and his Alpha Clock, a turret clock shaped like an 'A'. On display were some 16,000 objects gathered from 'the semi-barbarous nations of Asia and Africa[, ...] the rising states of the western world [...] and the vast colonial empire' as well as from the industrial regions of the United Kingdom. It was a celebration of peace and progress, and of Great Britain's manufacturing prowess.

Preparations for this 'millennial moment' had begun unpromisingly. The brainchild of Prince Albert attracted profound opposition from "soi-disant' fashionables and the most violent protectionists', as Queen Victoria described the project's conservative detractors. Elements of upper-class society resented the growing influence of the German-born prince consort, and wanted to preserve Rotten Row in Hyde Park for their horses. Reactionary politicians had feared that the working classes would take the train to descend upon London and unleash forces of anarchy. In previous decades, great assemblies of working people had signalled social strife and potential revolution. But from the grand opening ceremony of 1 May, when the monarch and more than 25,000 others crowded into the 19 acres of the Crystal Palace, and upwards of 700,000 people gathered peaceably in the park around it, the event proved an unexpected success. The rich and fashionable drew up in their carriages, and ordinary folk walked for miles for the privilege of passing through the turnstiles (made by Bramahs) into the magic, translucent kingdom. Fountains plashed, a gargantuan organ played and a vast choir sang. It was a celebration of peace and progress. By the time it drew to a close in mid-October, it had attracted more than 6 million visitors, drawn from all social classes and from round the world. At more than £500,000, profits were unexpectedly large, the revenues boosted by the innovative practice of charging a small fee for the privilege of visiting a public lavatory. Nearly 600,000 pennies were spent, accounting for the £2,427 taken for 'essential conveniences'. It was the defining moment of the vital, bustling, prosperous mid-Victorian age, or perhaps of any age, according to typical commentary. The *Illustrated London News* thought the Exhibition was 'the most instructive and memorable spectacle of our time, or of any time in the history of civilization'.[2] Adopting a cosmic perspective, *The Times* likened the opening ceremony to the Day

2 ILN, 31 May 1851.

of Judgment.[3] As expected, the British carried off more prizes for industrial excellence than any other nation, although there were impressive displays from the United States and European rivals. For more acute observers, there was evidence that Britain's leadership was open to challenge: particularly in the sensitive fields of locks and armaments.

* * *

For half a century until the summer of 1851, any person strolling down Piccadilly in London's West End would have seen a prominent display in the shop window of Messrs Bramah, locksmith and water closet manufacturer. Displayed in the window of number 124 Piccadilly was an unassuming padlock patented by Bramahs in 1784, and displayed in the shop window since before Joseph Bramah's death in 1814. The item, just 4 inches wide and 1.5 inches thick, was attached to a board with the following tantalizing words painted on it:

> *The Artist who can make an Instrument that will pick open this Lock shall receive 200 Guineas the Moment it is produced.*

The reward was enormous, a multiple of a working man's annual income, and the challenge apparently insurmountable. An ingenious mechanic had tried in 1817, but had retired in despair after a week's futile tinkering. For decades, the lock lived up to its manufacturer's proud boast that it was un-pickable, as by rights it should have been for all time: the arrangement of 18 separate sliders within a barrel 2.25 inches long and 1.5 inches in diameter gave rise to a staggering 470 million variations. Only the true key ought to have been able to open it. But come the summer of 1851, the attention of the world fixed on London as millions came to nearby Hyde Park to attend the Great Exhibition, a 38-year old American by the name of Alfred Charles Hobbs would have the temerity to take up the challenge.

In the United States, Hobbs had made a name for himself by opening supposedly impregnable locks, on one occasion winning $500 for springing a safe in the reading room of the New York Merchant Exchange, and on

---

3 'The progress of the human race, resulting from the labour of all men, ought to be the final object of the exertion of each individual[....] In promoting this end, we are accomplishing the will of the great and blessed God'. *Official Catalogue to the Great Exhibition of the Works of Industry of all Nations, 1851.*

another, breaking into the vast vault of the Bank of Morristown in New Jersey. He was mild-mannered and wore the clothes of a gentleman, but his delicate hands were capable of acts of criminal dexterity. In the full glare of the world's media, gathered to evaluate the host nation's claim to world domination in all matters relating to manufacturing, the American sought to replicate his feats, and thus prove the superiority of American locks. He was the representative of a New York firm called Day & Newell and its formidable 'Parautoptic' lock, designed with a hood over the key hole to prevent anyone seeing inside. The subsequent controversy played out against a backdrop of increasing nationalistic self-satisfaction.

One of the first firms visited by the Queen was Chubb & Co, Bramah's great rival in the British lock market. They 'really are wonderful, of every size and kind', she noted of the company's locks on 10 May. It was explained to her how these devices were impossible to pick – a fateful boast. And on 9 August, having rushed back from Osborne House to fit in another visit, she went 'to look at some very curious American locks', one of which was explained to her in detail by an American. 'It is very extraordinary but beyond my powers to attempt to explain,' she recorded.

It is tantalizing to think that the Queen might have met Hobbs and been bedazzled by his lock, but this seems unlikely, if only because the American would certainly have made a song and dance about any such encounter. If not the monarch, Hobbs did meet Charles Babbage, the inventor of the computer, who was impressed with 'a very curious lock of large dimensions with its internal mechanism fully exposed to view.' Hobbs gave Babbage, an expert on the mathematics of ciphering and deciphering, a 'very profound disquisition upon locks and the means of picking them, conveyed to me with the most unaffected simplicity'. Next time he came to the Exhibition, Babbage brought along the ageing Duke of Wellington, who 'was equally pleased with the lock and its inventor'. Babbage and Hobbs stayed in touch, the American explaining 'many difficult questions in the science of constructing and of picking locks'.[4]

At the Exhibition in early June, Hobbs declared that all the locks made in Great Britain up to that date 'admitted of being very easily picked'. He offered $1,000 to anyone who could pick his own Parautoptic lock. Having thus got the attention of a party of scientific gentleman, with a posse of reporters at hand, he produced two or three occult tools from his pocket and unceremoniously picked one of Chubb's patent detector locks. This act of mountebankish

---

4 Charles Babbage, *Passages from the Life of a Philosopher* (1864), 174.

American legerdemain took a mere 25 minutes. Hobbs subsequently proved as adept at manipulating the media, as the intricate components of locks and safes. Through a series of open letters, he declared his intention to pick the Bramah patent lock as well.

Before Hobbs could make his attempt, a committee of worthies was established, including Joseph Bazalgette, who went on to build London's system of sewers, and George Rennie, proprietor of the Albion ironworks in Southwark (and scion of a distinguished engineering dynasty). Under their close supervision, Hobbs was obliged to repeat his feat with the Chubb lock. Again, he produced two or three inoffensive-looking tools, and it was *open sesame* within minutes. On 22 July the arbitrators issued the following statement:

> We the undersigned hereby certify that we attended, with the permission of Mr. Bell, of No. 34 Great George-street, Westminster, an invitation sent to us by A. C. Hobbs, of the City of New York, to witness an attempt to open a lock throwing three bolts and having six tumblers, affixed to the iron door of a strong-room or vault, built for the depository of valuable papers[, …] that we severally witnessed the operation, which Mr. Hobbs commenced at 35 minutes past 11 o'clock A.M., and opened the lock within 25 minutes. Mr. Hobbs having been requested to lock it again with his instruments, accomplished it in the short space of 7 minutes, without the slightest injury to the lock or door. We minutely examined the lock and door (having previously had the assurance of Mr. Bell that the keys had never been accessible to Mr. Hobbs, he having had permission to examine the keyhole only). We found a plate on the back of the door with the following inscription: 'Chubb's New Patent (No. 261,461), St. Paul's Churchyard, London, Maker to Her Majesty.'

This outcome prompted a bilious response from Messrs Chubb, who lamented that they had not been present at Great George Street, and that moreover Hobbs had had access to the lock in question for weeks before the final test. The implication was that Hobbs was a cheat! Barely worthy of a response, retorted the American, as he protested the integrity of his actions. Letters to *The Times* were written. The balance of public opinion was that Chubb were behaving in an ungentlemanly, even an unmanly fashion, in not accepting that they were fairly beaten. In mitigation, it was noted that both Mr. Hobbs's parents had been English, and that he himself had escaped being born an Englishman by a whisper, his mother having emigrated to Boston while heavily pregnant. Under the circumstances, he was almost a gentleman.

The scene was set for the greater challenge. Following the arbitrators' instructions, the Bramah lock was removed from the shop window of the Piccadilly premises to an upper room. It was then enclosed in a block of wood and screwed to a door. The screws themselves were sealed and only the keyhole was accessible to Mr. Hobbs. When he was not at work, a band of iron was closed over the keyhole, sealed by Mr. Hobbs himself, to ensure that no one else could tamper with the device. 'The key was also sealed up, and was not to be used until Mr. Hobbs had finished his operations,' it was determined. 'If Mr. Hobbs succeeded in picking, or opening the lock, the key was to be tried, and if it locked and unlocked the padlock, it should be considered a proof that Mr. Hobbs had not injured the lock, but had fairly picked or opened it, and was entitled to the 200 guineas.'

The American set to work on 24 July. The picking of the lock took him a total of 51 hours alone in the room, spread over 16 days There was procedural wrangling, and for a time the work had to be suspended, but on 23 August, Hobbs displayed the Bramah lock opened to the arbitrators. On 29 August, he locked and unlocked the padlock in their presence. On the following day, they themselves locked and unlocked the padlock using the key, thus proving that Hobbs had satisfied the terms of the challenge and not inflicted any serious damage. The American then displayed the tools of his trade, including so-called 'thieves' wax', used to make an impression of the inner workings of the lock, a hinged mirror trained on the keyhole, a strong light and a variety of fixed and movable instruments. On 2 September, the judges declared that Hobbs had won the contest fairly, and the money should be paid over. Messrs Bramah did pay over the reward, but not before examining the innards of the violated lock and huffily telling the world that Hobbs had applied such force that the tumblers or slides had been bent backwards and forwards, and portions were almost filed straight through! This was scandalous and vulgar, entirely appropriate for an American, although within the letter of the rules. They avowed that the relevant components had been made out of iron rather than the steel now commonplace in English manufactures, and that they would modify the lock accordingly. Within two weeks, a new version of the lock was back in the window of their Piccadilly showroom and the reward was offered once again. There is no record of anyone rising to this new challenge.

Days before the conclusion to the Great Exhibition, the American thus dealt a blow to Britain's manufacturing pride. There were sinister undertones, with some commentators somewhat clumsily identifying consequences for Britain's national security. But given how long Hobbs had been at work, the contest did nothing to damage the reputation of the Bramah padlock for practical impregnability; if anything, it had enhanced the brand. 'The facilities given to [Hobbs] were such as no thief could ever possess,' commented *The Times*, 'even if he had the necessary ability, and it is quite clear that the operation has not been one of ordinary picking.'[5] A lock that could hold out for 16 days under such circumstances was more than adequate for the commercial market. The contest did no harm to Hobbs, either. He found himself lionized in lock-manufacturing circles and decided to stay in the United Kingdom. His invincible Parautoptic lock was awarded a prize medal by the Exhibition jurors, but it proved too expensive for the United Kingdom market. He abandoned it and his American employer, determining to set up a business in the United Kingdom. Like a star of the popular stage, he travelled the provinces to test his skills against any contender and added to his fame by publishing a short and highly readable account of the 'Lock Controversy'.

Those who examined the inner workings of the Bramah lock in the late summer of 1851 were amazed at the craftsmanship. It was a work of remarkable precision and detail. The barrel enclosing the mechanism was a mere 2.25 inches in length and 1.5 inches in diameter. 'The small space in which the works were confined, and its snug, compact appearance was matter of astonishment to all present.'[6] The lock was no more than half-a-century old, but it harked back if not to a pre-industrial age, but to a much earlier phase of the Industrial Revolution.[7] It was one of the very first complicated metal items made by machine. Buttons, nails and other simple objects had been knocked out for decades, but the lock presaged decades of engineering invention culminating in the mechanical plenitude of the Great Exhibition. Of course, in 1851, few would have understood the irony that the Bramah Lock – seemingly an amusing sideshow in the year of great achievements for industrial Great Britain – had been made by the man more than anyone responsible for

---

5 *The Times*, 4 September 1851.

6 Alfred Hobbs, *Rudimentary Treatise on the Construction of Locks* (1853), 127.

7 'As originally patented, the end of the key, of the pipe kind, has a number of notches or slots; usually six, of varying depth and, by means of these, corresponding sliders arranged radially in slots in a barrel are depressed against the action of a spring so as to arrive at a predetermined surface; this allows the key to turn the barrel round and enables the latter to shoot the bolt by a crank pin. The bit on the key merely determines the depth to which the key is to be pushed in.' 'Joseph Bramah and His Inventions,' by H. W. Dickinson, *TNS*, vol. 22 (1941–42), cited in L. T. C. Rolt, *Tools for the Job: A Short History of Machine Tools* (1965), 83–84.

the country's mechanical supremacy. Designed by Bramah, the machine used
to build the lock had been constructed by Henry Maudslay.

* * *

The impact of Hobbs and his locks notwithstanding, the American section
of the Great Exhibition threatened at first to be underwhelming. Unlike
for all other exhibitors, there was no government support for the country's
industrialists to ship over their goods, so they had to pay their own way or
plead for funds from George Peabody, a rich American banker living in
London. At the start, only 12,800 square feet of the allotted 40,000 was
occupied, with a result that the section looked like an empty prairie, according
to one satirical newspaper account. *Punch* suggested the sparse exhibits should
be shoved together, the underwhelming Cincinnati pickles piled up on top of
the decidedly ordinary Virginia honey, so that the space could be turned into
a sort of hotel for weary spectators. France, Germany and Austria, made a
much greater impression, their sections teeming with luxury goods, machinery
and armaments, commercial rivalry supplanting for a moment the permanent
political tensions among Europe's great powers. However, the Americans did,
in the end, make an impact away from Hyde Park, with Cyrus McCormick's
grain-reaping machine beating the domestic competition in trials in Essex,
while the yacht *America* roundly beat the British *Titania* off the Isle of Wight
in a race that was subsequently renamed the Americas Cup. This sporting
defeat signalled that the United States was now a naval power to be reckoned
with. Adding symbolic insult to injury, the British yacht was designed by the
great engineer, Robert Stephenson, himself. Above all, however, it was Samuel
Colt's revolvers that grabbed the attention of the complacent English.

'However profound may be the adjacent solitudes, here, at least, a knot of
enterprising travelers may always be seen gathered around a kind of military
trophy […] affixed to the northern side of the nave', noted *The Times*.[8] Mounted
on the wall was a display of numerous pistols of different dimensions. The
guns were small, light, accurate, and very deadly, as a series of trials made
clear during the summer of 1851. There were two basic models: one weighing
just over two pounds, the other about twice as much. They were light enough
to be stuffed into a belt or holster. These guns were deployed with murderous
effect in wars with the Seminole people in Florida, or against the Mexicans

---

8 *The Times*, 27 May 1851.

by the Texas Rangers, and by frontiersmen in their everyday skirmishes with inhospitable natives, operated on the principle that the cylinder housing six bullets revolved round with each bullet fired, until all bullets were exhausted. It was simple, but no one in England had seen anything like it before, and there was great curiosity as to how it worked. As a fascinated Charles Dickens explained:

> The revolving cylinder, behind the fixed barrel, is drilled with six holes, into which, one after another, the powder is rapidly dropped without being measured[; …] six balls are then taken in the hand, and also placed, one after another, in the holes. These balls are of conical shape [...] and are made of soft lead[. …] On pulling back the hammer with the thumb, after firing, the cylinder revolves one-sixth of its circumference, instantly bringing another hole, with its charge, in a line with the barrel. The barrel being rifled, and the charges in the breech airtight, none of the force of the powder is lost; and the balls are carried further, and with far greater precision than from an ordinary musket.[9]

The result was what *The Times* called 'terrible celerity,' a speed of deployment that put conventional double-barrelled weapons to shame. Eight British marines managed to fire 184 balls in five minutes. The efficiency of a body of cavalry was tripled by the use of such a gun: 100 men armed in such a way could do the job of 300 equipped in the conventional way. It was 'the most terrible instrument of destruction ever delivered into the human hand'. Even if it would be unsporting to place such a weapon in the hands of an English gentleman shooting snipe or pheasant, soldiers out on the frontiers of the empire in South Africa or India had to have it, immediately. 'We are bound to take care that our soldiers are supplied with the most efficient means of destroying an enemy,' *The Times* concluded. 'Colt's repeating pistols are the most efficient arm for mounted men and frontier troops, now known or used.'[10] To aid the sales effort, Colonel Colt made sure that many distinguished British soldiers were furnished with sample weapons when, after a summer in England, they returned to their postings in the far reaches of the empire. The pistol was self-evidently a quantum leap in the technology of warfare, comparable to the introduction of machine guns later in the century. The Duke of Wellington, no friend to technological innovation, 'was often observed at the exhibit

---

9 *Household Words*, vol. 9, 354.
10 *The Times*, 27 May 1851.

forcefully asserting the advantages of repeating firearms to an audience of officers and friends'.[11]

If politicians and soldiers were immediately eager to buy the guns, engineers had a more practical concern: they wanted to know precisely how Colt managed to make them. The American was thus invited to address the Institution of Civil Engineers over two evenings in late November 1851 to explain the practicalities, with Sir William Cubitt in the chair. This was a considerable event: the first time the prestigious Institution had been addressed by an American, and it was attended by high-ranking soldiers and politicians, eminent Americans and the flower of British engineering. Colt started by entertaining his audience with tales of efficacious slaughter of natives in Florida and Mexico, then he explained that eight-tenths of the cost of manufacturing the firearms was attributable to machines. According to the minutes of the meeting, Colt said he 'was induced gradually to use machinery to so great an extent, by finding that with hand labour it was not possible to obtain that amount of uniformity, or accuracy in the several parts, which is so desirable, and also because he could not otherwise get the number of arms made, at anything like the same cost, as by machinery.' In language that must have been music to the ears of Whitworth, the American explained that mechanization made it cheaper and quicker to produce identical, interchangeable parts in great quantities. He reportedly said, 'A duplicate can be supplied with greater accuracy and less expense, than could be done by the most skilful manual labour[. ...] On active service a number of complete arms may be readily made up from portions of broken ones, picked up after an action.'[12]

Whitworth, Nasmyth and Roberts had already implemented mass production in their own factories, but not on this scale. Pistols and locks caught the public and official imagination in a way that machine tools and engine components could not. Furthermore, since both Colt and Hobbs established their operations in central London, they were much easier for Charles Dickens and others to visit than were the factories of the industrial north. Samuel Colt's factory opened in Bessborough Place, Vauxhall Bridge Road, Pimlico, on 1 January 1853, on the bank of the Thames across the river from Maudslay's factory. Hobbs opened a three-storey factory in Cheapside, in the heart of

---

11 Cited by Nathan Rosenberg in his Introduction to *The American System of Manufactures* (1969), 15.

12 See Minutes of the *Proceedings of the Institution of Civil Engineers*, 'On the Application of Machinery to the Manufacture of Rotating Chambered-Breech Fire-arms', vol. 11, issue 1, January 1852, 30–50.

the City of London: 'The factory [...] may be distinguished at all times by the energetic snorting of a high-pressure steam engine, not a whiff of whose steam is allowed to escape without having completed its appointed tale of locks, bolts, bars, keys, screws, etc.'[13] Hobbs's own lucid explanation of the process of mass production at his London factory is worth citing at some length:

> If we suppose that a lock of particular construction comprises twenty screws and small pieces of metal, and that there are required [...] five sizes of such a lock; there would thus be a hundred pieces of metal required for the series, each one differing, either in shape or size, from every one of the others. Now on the factory or manufacturing system, as compared with the handicraft system, forging, drawing, casting, stamping and punching, would supersede much of the filing; the drilling machine would supersede the hand-worked tools. This would be done – not merely because the work could be accomplished more quickly or more cheaply – but because an accuracy of adjustment would be attained, so as no hand-work could equal [...] the advantage of the machine or factory mode of producing such articles is this, that they can be made in large numbers at one time, whenever the steam-engine is at work; and that when so made, the pieces are shaped so exactly alike, the screws have threads so identical, and the holes are bored so equal in diameter, that any one of a hundred copies would act precisely like all the others.[14]

Joseph Whitworth could not have put it better.

Visitors were amazed at the near full mechanization of the production process, which contrasted with the old-fashioned hand labour still prevalent in Wolverhampton, the traditional home of the British lock industry. The Colt factory in Pimlico inspired similar wonderment, even from James Nasmyth, who found there 'perfection and economy such as I have never seen before'.

* * *

In early October 1851, just a week before the spectacle of the Great Exhibition came to an end, Queen Victoria and Prince Albert travelled by train to Manchester and Liverpool. Despite the pouring rain, tens of thousands turned out to cheer their monarch and her consort. The people were wet and

---

13 'Hobbs' Lock manufactory,' *Engineer*, 18 March 1859, 188.
14 Cited by Rosenberg, op. cit., 13.

dirty, she noted after a day in Liverpool, but still the streets were crowded with well-wishers. The Queen and Prince Albert boarded the steamship *Fairy* and travelled along the enormous expanse of docks to the mouth of the Mersey and beyond. 'The mass of shipping is quite enormous, & forests of masts are to be seen,' the monarch wrote in her diary that night. On their return, the rain stopped briefly, and the royal party drove to the station in an open carriage, then took a train for 45 minutes to Patricroft, near Eccles, in Lancashire, where the aged Duke of Wellington and other grandees were waiting to escort her onto a barge for a short journey along the Bridgewater Canal.

For a few minutes, the modern, industrial world was left behind as the *Fairy* glided down the canal, the banks lined with cheering crowds. The Queen alighted at Worsley Park, the stately neo-Elizabethan residence of Lord and Lady Ellesmere where, after a cup of tea, she soon felt restored after the rigours of the day and ready for a dinner party. She sat down at 8 p.m. with the Ellesmeres, the Duke of Wellington, the Duke and Duchess of Norfolk, the great Cheshire magnates Lords Derby and Westminster – and James Hall Nasmyth, now proprietor of one of the most successful industrial enterprises in Victorian England. '[Nasmyth is] a Scotchman, & an engineer, a most intelligent man, who has entirely raised himself by his own efforts, & has a large factory at Paticroft [sic], where he has a steam hammer,' the Queen wrote that night, not entirely accurately, but clearly very much taken by him. 'He has made many wonderful & useful inventions, & is very simple, modest & unaffected. He is quite an enthusiast about the moon.' After dinner, Nasmyth enthralled his sovereign with a display of his drawings of the surface of the moon, which showed immense volcanic mountains and craters. 'It was extraordinarily interesting to hear what he had to say & he explained all so delightfully.'[15] In his *Autobiography*, published in 1883, Nasmyth laid great store by his family's aristocratic connections, and he would probably have been horrified to hear that the monarch, however impressed by his sketches of the moon, marked him down as entirely self-made. Within months of this great social triumph, Nasmyth was plunged into controversy and, indeed, the harmonious relations between the classes, so very much celebrated at the Great Exhibition, gave way to strife and contention.

---

15 Queen Victoria's diary for 9 October 1851.

CHAPTER 12

CAPITAL VS. LABOUR – THE GREAT LOCKOUT OF 1852

The Great Lockout of January to March 1852, which pitted master against operative, factory owners against employees, was 'carried on and fought out with the greatest vehemence and stubbornness, and excited the deepest interest through the whole country'. So wrote Thomas Hughes (later to be celebrated as the author of *Tom Brown's Schooldays*), in a humane account of this confrontation between capital and labour.[1] The Amalgamated Society of Engineers, Machinists, Millwrights, Smiths, and Pattern Makers (ASE for short) was formed in 1850–51, demanding that 'every machine must have a Union man to superintend it, and that he must be paid the full Union regulation wage'. Like modern teachers' unions objecting to unqualified staff working in the classroom, the ASE wanted to impose a strict limit on the ratio of apprentices to journeymen, and required all apprentices to serve five years before the age of 21. The union hoped that this would put an end to the practice of hiring qualified, older men as apprentices, and keeping them on low wages, and of hiring boys to do the job of men. It would also curb the employer's ability to add or decrease the size of the workforce in response to the fluctuations of the market. Such strictures were wholly unacceptable to Nasmyth, and indeed to most Manchester and London factory owners.

---

1 Thomas Hughes, *An Account of the Lock-out of Engineers 1851–52* (Cambridge, 1860) 2. There were two previous strikes of significance. The first of these took place in the winter of 1836–37, the second, in 1842, when the firm was caught up in the Great Plug Plot Riots, so-called because rather than smash machinery, the striking operatives realized they could bring production to a halt simply by removing the plugs from boilers and causing them to drain dry.

Less than two decades after self-acting machine tools had become commonplace, British engineering had become capital- rather than labour-intensive.[2] Businesses had become larger, and more dependent on expensive equipment and less on an aristocracy of skilled labour. As machine tools became universally available, the structure of the engineering industry was transformed, and the all-round expertise of the traditional millwright diminished in importance. Some jobs remained highly skilled and could not be mechanized, for example fitting and pattern-making, while tasks likes drilling, boring, turning and so forth could more easily be supplanted by the self-acting machines. Many factory jobs now required a man to look after a machine rather than exercising independent judgment and skill. 'One great feature of our modern mechanical employment has been the introduction of self-acting tools, by which brute force is set aside, and the eye and the intellect of the workman are called into play,' Nasmyth explained, many years later. 'All that the mechanic has to do now, and which any boy or lad of 14 or 15 is quite able to do, is to sharpen his tool, place it in the machine in connexion with the work, and set on the self-acting motion, and then nine tenths of his time is spent in mere superintendence, not in labouring, but in watching the delicate and beautiful operations of the machine.'[3]

Automation was not at all to the detriment of the working classes, he contended, because men's minds were improved by mere association with the beautiful geometric forms of the most modern machines: '[S]ome of [which] are exquisite in their structure and results as the finest astronomical instruments, requiring extreme accuracy and delicacy, and affording exercise for all the intellectual functions to meet the conditions constantly arising in dealing with the work, and that in itself is a most valuable kind of education'. Workers could be paid more, too, in proportion as they had time to look after more machines, as for example the labourer in his factory who was given six boring machines (which made the cylinders for steam engines) to attend to, and paid 1s for each. Better than reading an old newspaper and applying a drop of oil now and again, Nasmyth believed. The job was becoming so undemanding, that sooner or later steam engines would be made by girls. He said that in due course a

---

2 See Keith Burgess, 'Technological Change and the 1852 Lock Out,' *The International Review of Social History* (1969, vol. 14.), 215–36. Burgess shows how the firms involved in the Lockout were precisely those who had invested heavily in machine tools over the period 1835–50.

3 Report on Royal Commission on Trades Unions etc 1868, Q 19137, 19139, 19222 and 19299. See Nasmyth's evidence for Tuesday 14 July 1868.

weapons factory, for example, could operate without any employees at all: even Whitworth, apostle of automation, thought that was going too far.

Nasmyth was a proprietor whom skilled mechanics would respect: despite his gentle origins, he too was a worker with his hands, one who had built his enterprise through his own efforts. However, his philosophy as an employer took Manchester Liberalism to extremes. He believed in Free Trade in brain labour: quite simply, that his workers should be free to come or go, as they wished. That a third party like a union would be involved in setting the terms of their contract, he held to be 'derogatory to the interests of society' as well as noisome to the operations of his business, especially at time when demand for skilled labour outstripped supply and put the masters in an unfair position *vis à vis* the men. He would impose no conditions on his employees, in his mind a major concession, as the very best could find a ready market elsewhere at any time; they would stay with him if they wanted to, and he would strive to make it worth their while. 'There was the most perfect freedom between employer and employed,' Nasmyth explained. 'The only bond of union between us was mutual interest. The best [...] remained in our service because they knew our work and were pleased with the surroundings; while we on our part were always desirous of retaining the men we had trained, because we knew we could depend on them. Nothing could have been more satisfactory than the manner in which this system worked (Figure 7).'[4]

In the same vein, Nasmyth was adamantly opposed to traditional apprenticeships, whereby youngsters were indentured, often in return for paying over a premium, for a period of seven years before being anointed as journeymen, the next rung on the professional ladder. 'The fag-end of the feudal system,' is how he described this, as well as that it was downright disruptive, since apprentices typically behaved badly, were poor time-keepers, and could barely be disciplined. He would rather take on promising youngsters, 'intelligent well-conducted young lads', often the sons of workers already in his employ, and then promote them speedily if they were any good. The quick-witted ones could advance very rapidly, and be paid accordingly. 'If a clever lad in three years could do as much work and as well as a seven years' workman I would give him the same wages.' It took just three months to get an untrained mechanic up to the requisite level of skill, rather than the years required by a traditional apprenticeship. So he got rid of expensive adults, replacing them with young lads. Alternatively, he would hire experienced older men and

---

4 Nasmyth, *Autobiography*, 219.

FIGURE 7 Portrait photograph of James Nasmyth in later life (Courtesy of
Science Museum, London)

simply call them apprentices, meaning they were paid less than other men of
the same age and skill. 'Instead of having the old proportion of one boy to four
mechanics, I had four boys to one mechanic nearly,' he bragged. 'There were
an immense amount of labourers in the neighbourhood [...] and I got them
into my employment, and in a short time they were as good workmen as could
be desired.'[5]

There are echoes of Samuel Bentham's disregard for traditional norms of
behaviour here: Nasmyth operated a rational and meritocratic system that
paid no regard to the broader interests of a workforce in the process of coming
to terms with increasing mechanization. But Nasmyth paid well, enough
to ensure that his favoured employees did not have cause to 'look over the
hedge' for opportunities elsewhere, and there was the opportunity to climb

---

5 RC Evidence, 1868.

the career ladder from being a mere operative to a foreman, where one held sway over an entire department and would be eligible for company housing and a productivity bonus. Of course, hours were long: men were expected to work 12–14 hours a day, six days a week, 52 weeks a year, with holidays only at New Year and Whitsuntide. And conditions were dangerous, as dramatically illustrated a few years before the lockout, when a boiler blew up and propelled a three-ton steam hammer 43 yards into the air and across the works to land on a barge on the canal (which immediately sank), while multiple buildings were swept away like so many 'ninepins by a bowl', and the 65-foot high engine chimney fell down in a heap. John Rogers, the engine man, was killed outright, 'having been dreadfully burned and scalded', while the unfortunate Joseph Blears, sitting in the dressing shed some twenty yards from the boiler, was struck near the groin by a lump of iron. Henry Davis, also in the dressing shed, was hit by a cylindrical mass of iron weighing 7 cwt, which left his thigh severely lacerated, a piece of his thigh bone breaking away the wound. He died later, his widow receiving a munificent donation of 5.5 shillings, making up his wages to £1. Another man was hit in the back, while Thomas Hurst, hammer man at the forge, was scalded and buried under a pile of bricks. John Hepworth, foreman of the boilermakers, had a narrow escape as he, his wife and children had just left their cottage at the time of the explosion and were busy weeding the garden when a shower of bricks smashed the front windows of their house and pulverized the furniture inside. One of his sons was struck on his head by a two-foot-long slate tile, leaving the boy with a fractured skull.[6]

The immediate cause of the lockout was not the dangerous working conditions, which employees simply had to put up with, but resulted from a strike in late 1851 at Hibbert & Platt of Oldham, one of Lancashire's largest manufacturers of textile machinery, with 1,500 employees, whose equipment had been singled out for praise by the Queen when she had visited the Great Exhibition only months before. The union targeted the firm because of its practice of paying for piecework and overtime, and employing non-unionized labour. For employers, the strike was a provocation too far, and Nasmyth

---

6 The Coroner's Enquiry later established that the accident transpired following a trivial labour dispute, when a boy had refused to do what he was told by the deceased Mr. Rogers, and had thrown a piece of coal at the latter's head. The boy was dismissed, but Rogers had evidently been left in a semi-concussed state and he had allowed the boiler to overheat. Holbrook Gaskell testified that the man was a steady, sober employee, who had worked with the firm since its inception. The verdict: Accidental Death. The explosion caused more than £2,000 of damage. See the *Manchester Guardian*, 18 June 1845.

played a prominent role at a meeting at the Clarence Hotel, London, on Christmas Eve, in which he moved the motion that the Lancashire firms should join the national association of engineering employers in facing down the union. 'A general desire and determination was expressed that as the men had threatened a strike, their object and intention should be tested, and that the question between the employers and the employed, having been raised by the latter, should now be brought to some clear and definitive issue.'[7] It was all about showing once and for all who was boss: the employers refused to submit to the tyranny of the men, and wanted to demonstrate that they were *masters rather than slaves.* The phrase is picked up in *North and South,* the novel by Nasmyth's friend, Elizabeth Gaskell, which is set during a similar lockout:

> [T]he truth is, [the workers] want to be masters[, says Mrs Thornton, mother of John Thornton, mill-owner], and make the masters into slaves on their own ground. They're always trying it [...] and every five or six years, there comes a struggle between masters and men. They'll find themselves mistaken this time, I fancy, a little out of the reckoning. If they turn out, they mayn't find it so easy to go in again. I believe, the masters have a thing or two in their heads which will teach the men not to strike again in a hurry.[8]

Mrs. Thornton was right: in real life, the employers did have a trick up their sleeve, which was to shut down their factories and refuse to allow anyone to work unless they signed a declaration renouncing trade union membership for life. The subsequent lockout took effect from 10 January 1852, when Nasmyth, in common with most engineering firms of Lancashire and London, including Whitworth, Sharp, Roberts, Fairbairn and Maudslay, shut his doors to those of his workforce who belonged to unions, and in effect to all others besides. In total, some 23,000 men were thrown out of work in Lancashire and London, of whom only 7,000 were members of the union.

'The real grievance of the men is [...] that they are compelled to submit to hours of work so long that it is destructive of proper and reasonable enjoyment of life,' explained the union, somewhat plaintively. Overtime was acceptable, even desirable, when a by-product of unexpectedly buoyant orders, but not when it became a punishing routine. Forced to work until nine every night, the

---

7 *London Daily News,* 1 January 1852.
8 Elizabeth Gaskell, *North and South* (1854–55), Penguin edition, 138.

men become 'exhausted, jaded, wasted'. Many sought consolation in drink, and what we might call quality of life was impacted severely. Piecework was a means of paying men by results, rather than for the time they worked: an innovation that was fine in theory, but in practice had become 'the mode by which foremen ascertains the utmost that can be got out of the bone and muscle of each man in the establishment'. This then became the quantity expected of every man, every single day.[9] This was the rational, if unpleasant, response of entrepreneurs faced with the complexity of managing the large-scale industrial operations that had become increasingly commonplace by mid-century: no longer could owners rely on the quiet authority enjoyed by Henry Maudslay as he made his Sunday-morning rounds of the factory. Instead, they delegated production targets to the foremen, essentially middlemen who were powerfully incentivized to complete orders, no matter how high the human cost. As solidarity between employer and employed broke down, the seeds of Britain's notoriously poor industrial relations were sown.

According to Nasmyth's philosophy, those who went on strike, or even joined a union, exercised their right to leave his employment. Under such circumstances, it would not occur to him to deny them their freedom to leave, and thus to starve, or to interfere with the exercise of their free will by offering them their jobs back. Those workers who chose not to work were kicked out of their company cottages with three days' notice. Some 300 men were turned away when they reported to work in the first week of January. On a previous occasion, he had replaced the strikers with 'sixty-four first-rate men' shipped down from Edinburgh, together with their families and furniture. He deployed the same tactics in 1852:

PERFECT FREEDOM TO INDUSTRY AND TALENT[, *he advertised in Scottish newspapers at the height of the strike*]. ENGINEERS, MECHANICS, SMITHS, JOINERS AND OTHERS, desirous to bring their ability and labour to a GOOD and FREE Market, will find Employment, and liberal encouragement, by applying to Messrs JAMES NASMYTH & CO.[10]

Again, the strike was 'scotched,' suggesting some parallels with the way the fictional John Thornton in *North and South* broke the strike at his mill by importing Irish labourers. Until the great lockout, Nasmyth was friends with

---

9 *Observer*, 26 January 1852.
10 *Dundee, Perth, and Cupar Advertiser*, 17 February 1852.

the Gaskells, and may well have influenced aspects of the characterization, but there is no way that the Scot was a direct model for Gaskell's sympathetic character. For a start, Thornton had a sense of fundamental human solidarity with the workers, which Nasmyth simply did not: by this time, he was intensely cynical about the working classes, and unions in particular. Workers 'had a sad habit of getting drunk and coming to work on Wednesdays instead of Monday morning', he wrote. Machines, by contrast, 'never got drunk, their hands never shook from excess, they were never absent from work, they did not strike for wages [and] they were unfailing in their accuracy and regularity'.[11]

Those locked out or on strike were, at least initially, supported by a levy of one day's wages a week from those who remained in work, while those who crossed the picket lines were denounced as knobsticks and blacklegs. Despite the support of the more enlightened elements of the press, their cause was hopeless in the face of employers' hardball tactics. Financial support dwindled, and the manufacturing communities of the North suffered great hardship, such as the clemming – or starvation – described in Gaskell's novel, and the strike collapsed after three months, the men forced back to work without gaining any concessions. In the novel, Thornton's business goes bankrupt following the lockout, and is only rescued through the generosity of Margaret Hale. By contrast, Nasmyth's business in fact went from strength to strength. 'I placed myself in an almost impregnable position,' he bragged in his *Autobiography*, 'and showed that I could conduct my business with full activity and increasing prosperity.'[12] His claim to have done this while maintaining good feeling between employed and employer, was far-fetched, however. Those who came down from Scotland to relieve the strike were blacklisted and could not find work anywhere else in the United Kingdom if they left Nasmyth's employment. Following the lockout, he stepped up the use of self-acting machines, telling the Royal Commission on Trades Unions (1868) that in this way he had managed to reduce his labour force by 'fully one-half'. He insisted he could have created hundreds more jobs, but for the troubles associated with unions.

For all his sang-froid, Nasmyth was deeply affected by the conflict. 'I have never been so pulled about to all quarters of the compass as I have this winter on business affairs,' he wrote in March 1851.[13] He suffered from exhaustion and the Victorian equivalent of stress: having been a pioneering entrepreneur, engineer and industrialist, the challenge of running a large complicated

---

11 Nasmyth, *Autobiography*, 164.
12 Nasmyth, *Autobiography*, 299.
13 Letter to George Loch, cited in Musson and Robinson, op. cit., 508.

operation had come to overwhelm him. The pressure of work left no time for friends or hobbies, he lamented, but he consoled himself with the thought that the hard work was all in a good cause, bringing him nearer to the cottage in Kent that he dreamt of. As he later explained, having decided to retire and invest his fortune in 3 per cent consols, government securities,

> I was so annoyed with walking on the surface of this continually threatening trade union volcano [...] that I was very glad to give it up and retire from the business [...] at least ten years before the age at which I would otherwise have retired[....] I heard the distant rolling of thunder, and saw that there was a storm coming on, so I made a compromise between my desire to be very rich and my desire to be very tranquil and comfortable, and resigned the business into the hands of my sleeping partner, and appointed a practical manager to take my place.[14]

In reality, Nasmyth was already very rich, and he reached the decision to cash out and retire from business 'and enjoy the residue of my life in the active pursuit of all my favourite occupations and intercourse with such dear old friends as are left'.[15] Nasmyth finally sold the business in 1856, retiring not to a cottage, but to a mansion with 16 acres near Penshurst in rural Kent, about as far geographically and culturally from the industrial North as it is possible to be. Remote from the scene of so much strife, he reflected that on balance the constant turmoil in labour relations was useful for society at large, as it forced the pace of labour-saving innovations. 'Such has been the stimulus applied to ingenuity by the intolerable annoyance resulting from strikes and lock-outs,' he said in 1868, 'that it has developed more than anything those wonderful improvements in automaton machinery that produce you a window frame or the piston rod of a steam engine of such accuracy that would make Euclid's mouth water to look at.'

* * *

Nasmyth was not the only one considering retirement at the time of the lockout: on 24 January 1852, less than a month after the job action started, the *Manchester Guardian* carried an advertisement for the sale of Richard Roberts's

---

14 RC Report 1868, Q's 19222 and 19299.
15 As he explained to Smiles in 1882, while the proofs for the *Autobiography* were being prepared.

Globe Works. The entirety of the factory's tools and stock in trade was to be auctioned off, including: lathes, slide rests, screw cutting engines, immense planing machines (one with a 40-foot long bed); drilling, boring and broaching machines, turret clocks, throstles, mules, anvils, hoists, patterns, cranes, boilers, a cast-iron billiard table, two locomotive boilers, a cigar-making machine and remnants of the great punching machine that had been instrumental in building the tubular bridge across the Menai Straits. Like Nasmyth, but with far more sympathy for the working man, Roberts had resolved to abandon the responsibilities of a large enterprise. He was to become a self-employed consultant engineer: a career pursued by many distinguished figures, not least Isambard Kingdom Brunel. For Roberts, it was a highly productive period: from the time of the dissolution of Sharp Roberts in 1843 until his death in 1864, Roberts took out 18 patents for objects as various as turret clocks, watches, traction engines, double-keeled ships with twin propellers (his patent of 1852 contained 33 improvements to existing technology), a floating lighthouse, technology for disposing of sewage and a prototype omnibus for use in London. In total, he took out 27 patents during his life, more than any other individual at the time. But without a business partner such as Thomas Sharp, Roberts struggled to make money: none of the inventions was commercially successful. The patent for the self-acting mule had run out in 1846, and he had no steady source of income. His resources were gradually depleted until eventually he moved to London, where he based himself in Adam Street, just south of the Strand. He lived there, looked after by his daughter, Eliza Mary, in increasing poverty and ill health.

In the aftermath of the Great Exhibition and the lockout, Joseph Whitworth, who was far from considering retirement, was sent by the government to the United States to assess its manufacturing capabilities. This was in belated recognition that the nation, however apparently supreme in manufacturing matters, had something to learn from the Americans after all, especially in the field of armaments. When Whitworth finally delivered his report in 1854, his assessment was finely balanced: on the one hand, machinery in the United States was generally less sophisticated than that in use in the United Kingdom. Engine or machine tools were 'similar to those in use in England some years ago, being much lighter than those in use now, and turning out less work in consequence'. But Americans were better at mechanization of light metal goods, for example, small arms and locks, and they had also recently invented the sewing machine, which was becoming generally available and transforming the clothing industry.

With his experience of the lockout clearly in mind, Whitworth reported that American workers were much more willing to embrace new technology than were their British peers. 'The workmen hail with satisfaction all mechanical improvements, the importance of which, as releasing them from the drudgery of unskilled labour, they are enabled by education to understand and appreciate,' he wrote. The supply of labour in the United Kingdom was comparatively abundant, and workers feared the progress of invention as being more likely to put them out of a job. This frustration was echoed by Alfred Hobbs, who was eventually to leave Britain, in part as a result of exasperation with the narrow-mindedness of the British working man. 'In America they might set to work to invent a machine, and all the workmen in the establishment would, if possible, lend a helping hand,' the locksmith wrote. 'If they saw any error they would mention it, and in every possible way they would aid in carrying out the idea.' (Hobbs returned to the United States in 1865, where he helped run a sewing-machine company and a firearms enterprise.)

In England, however, it was quite the reverse. 'If the workmen could do anything to make a machine go wrong they would do it[.... He thought] the great obstacle in the way of the gunmakers of Birmingham in introducing machinery, was the opposition of the workpeople to such innovations'.[16] (Birmingham was the traditional home of the light-arms trade, where components for guns were still made by hand.) Nasmyth corroborates this, contrasting the Americans' innovating energy with the 'traditional notions and attachment to old systems' widespread in England.[17]

Whitworth found that American operatives felt no compulsion to form unions as a way of fighting off mechanical progress. 'The principles which ought to regulate the relations between the employer and the employed seemed to be thoroughly understood and appreciated,' he declared. Capitalists were therefore encouraged to put their money to work, while 'the intelligent and educated artisan is left equally free to earn all he can, by making the best use of his hands, without let or hindrance by his fellows.'[18] Whitworth was as hostile to unions as any other British industrialist who had locked the doors of his factory in the winter of 1852. Where he continued to differ from hardliners,

---

16 *Journal of the Society of Arts* V, no. 219 (30 January 1857) 165.

17 Cited in Musson 500, also *Report of Select Committee on Small Arms, 1854*, as well as his *Autobiography*.

18 Joseph Whitworth, *Official Report on the New York Industrial Exhibition of 1853*, published in *Miscellaneous Papers on Mechanical Subjects*, 172–73.

such as Nasmyth, was in his sincere belief in the power of education to improve workers' lives. For now, however, philanthropy took second fiddle to the pursuit of a Holy Grail as elusive as that of true planes: the application of precision engineering to the manufacture of weapons. As John Bright, pacifist and fellow northern industrialist, mused when Whitworth came to visit him in London days before the United Kingdom entered the Crimean War in late March 1854: '[I]t is melancholy to think that the greatest ingenuity in the country is now employed on instruments of destruction'.[19]

---

19 John Bright's *Diary*, 17 March 1854.

CHAPTER 13

INSTRUMENTS OF DESTRUCTION

It was said, and not entirely in jest, that in the first half of the nineteenth century the British were more concerned to see improvements in sporting guns used by gentlemen to shoot snipe or partridge, than in the weapons used by ordinary soldiers to pursue the nation's vital interests in battle. 'Money and skill were bestowed without stint on a rifle to bring down a deer, or on a fowling piece with which a pheasant was to be shot,' noted a contemporary observer, 'but any weapon, however clumsy, was thought sufficiently good when the issue of a battle or the fate of an empire was in balance.'[1]

The smooth-bore muskets used at the time of the Napoleonic Wars and thereafter, were hopelessly inaccurate. Their barrels were knocked together by blacksmiths, and the contents of cartridges (a soft lead bullet and gunpowder) were rammed into the muzzle of the gun with a rod, the lead often losing its shape in the process. These weapons had low muzzle velocity, and the ball did not fly straight and true. At the Battle of Salamanca in July 1812, for example, 3,500,000 bullets were fired, but only 8,000 men fell (and some of those were hit by the 6,000 cannon balls fired, or mown down by cavalry charges); at Waterloo, it was observed that one side of a square of British infantry firing into a body of advancing French cavalry, was able to dislodge merely three or four Frenchmen from their saddles. At 450 yards, the soldiers were at a safe distance from all small arms. In the 1830s, the military undertook a series of tests with the service musket, which showed that a target 3 feet wide and 6 inches high could be hit consistently at a distance of no more than 150 yards; any further than that, the target had to be widened to six feet across. At 250 yards, not a single shot found its mark. Traditional muzzle-loaded muskets were so woefully inaccurate that soldiers were on one occasion ordered to fire 130 feet above a man, in order to have a chance of hitting him at 600 yards.

---

1 Tennent, op. cit., 4.

They were told to level their weapons, rather than aim them, in the general direction of the enemy. 'In other words, if you wish to hit a church-door, aim at the weather-cock.' At 200 yards, a trained marksman using a military-issue musket could not hit an 18-square-foot target more than 1 in 20 times. Bigger guns, such as the cannon used in pitched battles on land or at sea, were just as inaccurate at long ranges, and powerless to penetrate the ironclad armament increasingly carried by warships. And this was in an age when Whitworth could measure to a millionth of an inch, and Nasmyth's hammers could smash home with extraordinary power and precision.

As the Duke of Wellington justly observed: 'Looking at the amount of mechanical skill in the country, and the numerical weakness of our army as compared with those of the great continental powers, British troops ought to be the best armed soldiers in Europe.'[2] The Maudslay-inspired revolution in precision engineering had passed the army by. Only the display of weaponry from rival nations at the Great Exhibition jolted the military out of its official complacency, and in 1852 gunmakers were asked to submit designs for a more accurate and effective service weapon. The result was the Enfield rifle, manufactured at the Royal Small Arms Factory in the North London suburb of that name, an establishment that dated back to the Napoleonic Wars. With a barrel 39 inches long, and a bore of 0.577 inches, and weighing 14.5 lbs, the Enfield was a dramatic improvement on the 'Brown Bess' smoothbore musket (and its immediate successor, the Minié rifle). It would perform serviceably well in the Crimea, smiting the enemy 'like a destroying angel' (according to *The Times*) at the battles of Alma and Inkermann. Yet, even before this crucial test of Britain's imperial and military strength, a new generation of soldiers and politicians became concerned about the new weapon's very palpable defects. Lord Hardinge, the veteran soldier who succeeded Wellington as commander-in-chief of the British Army in September 1852, explained in a memorandum of 1854: '[O]ne rifle [from the Enfield factory] shoots well, another ill, and the eye of the best viewer can detect no difference in the gun to account for it'. How to account for these variations, given that all the rifles were manufactured apparently in the same way? It was a version of the question that had challenged Joseph Clement as he puzzled out how to manufacture components for the Difference Engine to a level of precision where they were interchangeable – or faced by Roberts, Nasmyth and Whitworth as they invented mass production, and which had apparently been answered by Samuel Colt in his revolver factory.

---

2 Tennent, op. cit., 11.

'The soul of manufacturing is repetition,' as Whitworth put it; somewhere in the Enfield manufacturing process were errors and flaws that made this ideal impossible to attain. In 1853, the government commissioned the Enquiry into Small Arms Manufacturing, putting James Nasmyth in charge. Alert to his patriotic duty and to the fascinating mechanical challenges, Whitworth also started to investigate the problem. The day after he met John Bright, he was one of three expert witnesses called to the House of Commons to field questions on small-arms manufacture. He was asked by the government to provide a complete set of designs for new machinery at Enfield, potentially a lucrative assignment as the government was minded to order a million rifles to meet the demands of war. Whitworth spurned this substantial commercial opportunity, as he 'felt it inconsistent with his reputation to make machinery for the multiplication of an imperfect article on the vague chance of correcting errors, the precise nature of which had not yet been discerned.'[3] The order was placed instead with the American firms Robbins & Lawrence (which had exhibited alongside Colt at the Great Exhibition) and the American Manufacturing Company, signalling the achievements of the US machine-tool industry.

Whitworth declared himself willing to invest his personal time, effort and money to investigate the mechanical properties of rifled weapons and their ammunition, which had never been subject to thorough scientific investigation. He would give his time *gratis*, but he requested that the government provide the funds to erect a shooting gallery on his Manchester estate. There was resistance from elements of the military establishment, adamant that it was pointless indulging in expensive experimentation at the time of war, just as reactionaries had fought the development of the block factory in a previous generation. But Hardinge supported Whitworth, arguing that if there was a secret to the manufacture of rifles, only this engineer could discover it. ('No gun-smith could imitate the most perfect rifle, nor does he know why it shoots well or ill; but if the secret is to be discovered, it may be copied by machinery, and Mr Whitworth is very confident that he can discover and can copy it.') In mid-1854, Whitworth was provided with the funds (some £16,000 in total) to build his gallery, a brick tunnel 500 yards long, 16 feet across and 20 feet high, with a tiled roof and windows along its south side. The gallery was ready in September, but blew down in a storm immediately afterwards, so it was not until March 1855 that experiments began in earnest. Some ninety years before stroboscopes were used to take photographs of bullets moving at more than

---

3 Tennent, op. cit., 30.

the speed of sound, Whitworth embarked on a series of experiments that led to the first theoretical and practical understanding of the science of projectiles. As the eminent scientist Professor Tyndall said, 'his genius brought scientific orderliness to what hitherto had been the most speculative experiments'.

His work was too late to influence the course of the Crimean War, the conduct of which exposed the lamentable inefficiencies of the British ruling class, notwithstanding the heroism of the Charge of the Light Brigade and the saintly efforts of Florence Nightingale. Whitworth's friend, John Bright, talked publicly of an Angel of Death stalking the land: '[Y]ou may almost hear the beating of his wings,' he told the House of Commons in late December 1854. Privately, the politician and businessman's rhetoric was less uplifting, as he bemoaned the impact of the war on the price of tallow, a crucial ingredient for his carpet factory. ('Our carpet trade grievously injured by war raising the price of tow [...] the [war] deranges trade, and deters everyone from contracting engagements.'[4]) There was some consolation that Whitworth manufacturing standards aided the war effort in other ways. As *The Times* later explained:

> The Crimean War began, and Sir Charles Napier demanded of the Admiralty 120 gunboats, each with engines of 60 horsepower, for the campaign of 1855 in the Baltic. There were just 90 days in which to meet this requisition, and, short as the time was, the building of the gunboats presented no difficulty. It was otherwise however with the engines, and the Admiralty were in despair. Suddenly [...] the late John Penn solved the difficulty[. ...] He had a pair of engines on hand of the exact size. He took them to pieces and he distributed the parts among the best machine shops in the country, telling each to make ninety sets exactly in all respects to the sample. The orders were executed with unfailing regularity, and he actually completed ninety sets of engines of 60 horsepower in ninety days – a feat which made the Continental Powers stare with wonder, and which was only possible because the Whitworth standards of measurement and of accuracy and finish were by that time thoroughly recognized and established through the country.[5]

The manufacture of components such as cranks and connecting shafts was thus farmed out to numerous factories. John Penn and Maudslay's were

---

4 John Bright's *Diary*, 15 April 1854.
5 *The Times*, 24 January 1887.

given the task of assembling the components to create the finished engines to be delivered to the Royal Navy. By the 1860s, 'the several portions of a most elaborate engine or machine may be made and finished by different workmen and at different places, and when brought together they fit and adjust themselves with a precision that leaves nothing to be wished for'.[6]

With the tenacity of purpose, meticulous attention to detail and patience that he had dedicated to other intractable problems, for example true planes or the perfect screw, Whitworth threw himself into the task of finding out how and why bullets fly in a straight line, hit their target, and can be made to do so time and time again. Abandoning the day-to-day running of his factory to professional managers, Whitworth personally supervised the experiments. The testing range was equipped with a target on wheels that could be moved backwards and forwards to facilitate shooting at different distances, rests for steadying the aim, and screens of light tissue paper set up every 30 yards to capture the trajectory of a bullet in flight. There were also bags of chaff placed at intervals that would stop the bullets so the degree of rotation could be measured. Before the shooting started, Whitworth's engineers fashioned trial barrels made out of segments of tubing held together by external hoops, then reheated the whole until it was red-hot and gave it a twist. It was known that a degree of spiralling in the barrel would help give a bullet spin and thus give it stability: but how much twist, and at what speed and with what force would the best results be guaranteed? (The Enfield took one turn round the interior of the barrel in 78 inches.) And, crucially, what sort of projectile would deliver unfailing accuracy, with minimum wear and tear on the barrel? Whitworth knew that a conical bullet would outperform a spherical ball – but in his first experiments with a long bullet gave poor results and Whitworth took up the idea of a hexagonal bullet in a hexagonally rifled barrel.

The end result, arrived at after two years of trial and error, consisted of the following elements: an 'improved system of rifling; a turn in the spiral four times greater than the Enfield rifle; a bore in diameter one-fifth less; an elongated projectile capable of a mechanical fit [...] and a more refined process of manufacture'.[7] As in the case of the self-acting lathe, the steam hammer or indeed the railway engine, there were parts of the system eventually patented by Whitworth that were not original: for example, Isambard Kingdom Brunel had toyed with the idea of a hexagonal barrel. But in common with Henry

---

6 Cited in Davenport-Hines *Capital, Entrepreneurs and Profits* (1990), 240.

7 Tennent, op. cit., 51.

Maudslay or indeed James Watt, it was Whitworth's ability to manufacture the new rifle, rather than merely design it, that was the real achievement. He developed new broaching and boring machines that meant that the new rifles and their snug-fitting bullets could be manufactured to an unprecedented degree of replicable accuracy.

The results were spectacular: the rifle was tried against the Enfield at Hythe, in April 1857, and it outclassed the official military weapon. The results of the shooting at 500 yards show the Whitworth bullets clustered in a radius of four and a half inches from the centre of the target – compared to a literally scattergun 27 inches for the Enfield. Whitworth's rifle retained its superiority at longer distances, indeed at 1,800 yards the trials were disbanded as the Enfield could not come anywhere near the target, while the Whitworth kept on hitting home. The bullets penetrated three times further through wood or sandbags than the Enfield. As *The Times* reported, the trials 'established beyond all doubt the great and decided superiority of Mr. Whitworth's invention. The Enfield rifle, which was considered so much better than any other, has been completely beaten. In accuracy of fire, in penetration, and in range, its rival (the Whitworth) excels to a degree which hardly leaves room for comparison.'[8]

George Bidder, the former human calculating engine, tested the rifle by firing 100 rounds on one day, 60 on the next, then 40 rounds, and so on for ten days, without cleaning it, and at the end of the trial it was working as well as on the first day. A more illustrious endorsement of its properties came on 2 July 1860, when Queen Victoria came to Wimbledon Common to open the National Rifle Association's annual rifle shooting match. After a rousing welcome from local grandees and hundreds of militia men, she entered a small tent to take her own shot. She pulled the trigger of one of his rifles and hit the target a half inch from the centre, at a range of 400 yards. 'The Rifle was fixed by Mr Whitworth himself,' the monarch wrote in her diary that night. 'I gently pulled a string & it went off, the bullet entering the bull's eye!!' She was presented with a medal. With pardonable hyperbole, this was described in the House of Commons as 'probably the most marvellous shot ever fired from a rifle'. Although the Queen merely gave a gentle tug on the string, the shot was the culmination of a lifetime's work: the rifle was attached to a steel slide riding on true plane surfaces, sliding on other true planes, ensuring that the recoil did not disturb the accuracy of the aim at the instant of firing. It was indeed marvellous that the accuracy was guaranteed, testament to the

---

8 *The Times*, 23 April 1857.

advances in precision engineering pioneered by Maudslay and taken forward by Whitworth.

Competitors criticized the Whitworth rifle for its expense, and the British Army thought its bore too small for combat, and could never bring itself to accept the practical utility of the weapon, despite its unquestionably superior performance. Indeed, in the American Civil War, it was found that its efficacy as a regular battlefield weapon was diminished as it did not work well with ordinary bullets: unlike in the case of screws, Whitworth was not able to promote a new global standard for bullets. But the rifles were greatly prized by snipers in the Confederate forces, used to pick off Union soldiers at great distances. 'I had no ammunition to spare and did not reply to the continual fire of the enemy except with five Whitworth rifles,' reported General Cleburne of the Army of Tennessee, describing the fight at Liberty Gap in 1863. '[These] appeared to do good service. Mounted men were struck at a distances ranging from 700 to 1,300 yards.'[9]

It was a similar story with Whitworth's field guns: the principles on which he had built his rifles worked perfectly well with heavy ordnance. His neighbours in Manchester may have been disconcerted as he commenced testing of his brass 24-pounder howitzers in his (admittedly extensive) back garden. In 1860, in search of a safer location, the trials moved to the sands of Southport, north of the Mersey. With 2.5 pounds of powder, fired at an elevation of 8.25 degrees, the gun sent a shot of 24 pounds a distance of 3,500 yards, or nearly two miles. This so far exceeded the ordinary range of a mere mile and a half, that the shot bounced off the beach, ricocheted to the right, and penetrated the window of a marine villa, rolling on to the carpet in front of retired lady who was sitting by her drawing-room fire. Though greatly astonished, the woman suffered no injury. Whitworth was seeking the same predictability and replicability of results for his larger guns as for his rifle and other machines. 'The target was placed one thousand yards off,' reported the *Mechanics Magazine*, 'it was six feet square and had a bull's eye two feet square. Whitworth fired ten shots. The first two passed close over the mark; but each of the succeeding shots were sent through the middle part of the target, two of them leaving their holes in the bull's-eye.'[10] On 25 September 1862, bigger guns were tested at Shoeburyness in Essex, where they penetrated four inches of armour at a distance of 600 yards. The velocity, accuracy and power of these new weapons were unprecedented

---

9 Cited in Joseph G. Bilby, *Civil War Firearms: Their Historical Background and Tactical Use* (Conshocken, Pennsylvania, 1996), 122.
10 Quoted Atkinson, op. cit., 60.

and pointed the way to a new form of warfare, where armies could pulverize each other from a distance, and ironclad warships could blow each other out of the water. But, as with the rifles, Whitworth's heavy guns did not find favour with the naval or military establishments, who opted for less-accurate products made by Sir William Armstrong, the great Tyneside engineer and entrepreneur. One reason was that Whitworth's gun barrels did, from time to time, burst when fired, testament to the limitations of cast iron as the raw material for increasingly powerful weapons: this was a commonplace risk for heavy weaponry of the day, but still incompatible with Whitworth's reputation or his aspirations. This failure would inspire Whitworth to begin, late in life, another round of experiments as he sought to develop steel as a viable material for large-scale cannonry. Another reason for official indifference to Whitworth's inventions was skulduggery: his rival, Armstrong, manoeuvred himself into a position as both advisor to the government and a leading supplier of armaments. Even in an age when concepts of conflict of interest were not well established, this caused mid-Victorian eyebrows to be raised. The rivalry between the two men was played out on shooting ranges, and in the press.

A third explanation for Whitworth's exclusion from official contracts, was undoubtedly his obstinate perfectionism: he simply would not compromise his designs to address what he considered to be the piffling and ill-informed concerns of committee men or army staffers. His technology was, however, taken up by Emperor Louis-Napoleon of France, who made him a member of the Légion d'honneur after seeing his field gun in action, and was used to great destructive effect by the Confederates in the American Civil War. Although frustrated that his inventions were not adopted widely at home, he remained busy and productive until well into his eighties, as we will see in the next and final chapter, which assesses the personal and collective legacies of Henry Maudslay, Joseph Whitworth and the Maudslay men.

CHAPTER 14

ENDINGS AND LEGACIES

The two most prominent engineers of the age died within weeks of each other in the autumn of 1859. Isambard Kingdom Brunel, who died on 15 September, and Robert Stephenson, on 12 October, ended their lives as wealthy, celebrated figures, some 3,000 people attending the latter's funeral at Westminster Abbey.

By contrast, the last years of the oldest surviving Maudslay man were spent in poverty and obscurity. Richard Roberts, perhaps the most ingenious mechanical engineer of all, was living with his daughter in modest rooms off the Strand, a victim not merely of his uncommercial disposition, but of the American Civil War. The conflict deprived northwest England of the raw material for its factories, producing a cotton famine that led to real starvation for thousands of workers. Roberts's remaining assets in Lancashire were rendered worthless. He died at home in London on 11 March 1864, at the age of 75, his inventive spirit undiminished despite his straitened circumstances. It was said that on his very last day he was working on designs for a slate-cutting machine and for improving a sewing machine.

Friends and admirers such as William Fairbairn and Lord Broughton had got wind of his straitened circumstances and had started to raise money with the intention of providing him with a pension. 'A considerable sum had been subscribed when the intelligence came that he was gone,' noted the *Engineer* in its obituary. The pension of £200 a year was diverted to his daughter. At noon on the day of his funeral, many leading engineers gathered at his rooms to pay tribute: those present included Henry Maudslay, a son of the master; Charles Beyer, by now an independent and hugely successful railway engineer; David Napier, the shipbuilder; William Muir, formerly of Maudslays and Whitworths; and John Ramsbottom of the Crewe works. They set off with the coffin to Kensal Green cemetery for the interment of this supremely gifted and greatly loved engineer.

* * *

Though never troubled by poverty, Charles Babbage ended his life, like Richard Roberts, a failure in the eyes of the world, and in his own estimation. After the Difference Engine, Babbage directed his efforts to the creation of an even more-complex machine, namely the Analytical Engine. This was conceived as a 'general purpose machine capable of calculating virtually any mathematical function'. Programmed with punched cards (a technique borrowed from the Jacquard loom), this machine had many of the features of the modern computer, including the distinction between memory (the store as he called it) and processing (the mill). Babbage pleaded his case with successive governments, who deemed the project worthless, and were entitled to be confused, as Babbage was 'in the habit of starting a model and then abandoning it in an unfinished state in order to start a new one.'[1] The machine was never built, but it was studied and appreciated by far-sighted contemporaries such as Luigi Federigo Menabrea, the engineer who became the Prime Minister of Piedmont, and Ada Augusta, countess of Lovelace (and Lord Byron's daughter). 'We may say most aptly that the Analytical Engine weaves algebraical patterns just as the Jacquard loom weaves flowers and leaves,' she wrote.[2] When Babbage died in late 1871, distracted to the last by the cacophony of noises reaching his sickroom from the street, he was considered something of an oddball. As the clerihew caught it:

Mr. Babbage

Lived entirely on cabbage.

He used his head, rather than his thumbs

In inventing his machine for doing sums

But of course he was staggeringly ahead of his time, his designs and musings on the science of computing anticipating the era of Google and Big Data. In 1991, the Difference Engine was finally built to Babbage's design by a team from the Science Museum; it worked perfectly. Ben Russell, Curator of Mechanical Engineering at the Science Museum, believes this machine could have been constructed in the 1830s, and that it was lack of financial resources rather than engineering expertise that meant it was not. Meanwhile, Joseph Clement, the man who gave physical expression to Babbage's first big idea, died at home in Southwark in February 1844, aged 65. Clement was

---

1 L. H. Buxton, 'Charles Babbage and His Difference Engines,' *TNS*, vol. 14 (1934): 51.

2 Babbage, *Passages*, 27.

unmarried but left £4,000 to a natural daughter; his tools, the subject of the dispute with Babbage and at one point the *ne plus ultra* of pre-Victorian engineering technology, were also valued at £4,000 and were left to a nephew who carried on in business for a short time.

* * *

In April 1871, Joseph Whitworth got married for the second time, to Mary Louisa née Broadhurst. This time, Whitworth married emphatically into Manchester's industrial aristocracy: his new bride's father, Daniel, was the city treasurer of Manchester, an influential position, and her elder brother was Henry Tootal Broadhurst, chairman of Tootal Broadhurst Lee, the leading textiles company of the day (and descendant of Nasmyth's patron, Edward Tootal). While keeping hold of The Firs (which was subsequently leased to C. P. Scott, the great editor of the *Manchester Guardian*), Whitworth and his second wife built a country house at Darley Dale, near Matlock in the Peak District, where he treated himself to all modern conveniences, including a marble bath with piping hot water, which much amused Jane Carlyle when she stayed overnight. He and his wife did not have children. They indulged in landscape gardening on a monumental scale, constructing a vast park at the site of an old quarry. Whitworth kept a herd of longhorns, trotted around the country lanes with a pony and trap, and played billiards on a snooker table with an impeccably true plane for a surface.

While such distractions were of a piece with his standing as a great industrialist, Whitworth never retired as such. In 1874, he converted his business into a limited liability company and became a pioneer of worker democracy, sharing control with 23 senior staff, and encouraging ordinary operatives to invest £25 in shares. Having perfected gun design, he moved on to steel manufacturing, where tremendous advances had been made by the inventor Charles Bessemer, who patented his process for mass-producing high-quality steel in 1856. This did not satisfy Whitworth, who instead pioneered a new method of making ductile steel. This involved applying extreme pressure to the metal when in a molten, fluid state, using a hydraulic press. In his eightieth year, he sold the factory in central Manchester and opened a steel-works and machine shop at Openshaw on the outskirts of the city. In 1883, he displayed some of his steel forgings at the Great Exhibition in Paris, including a heavy shaft for a screw propeller and steam cylinders for a steamship. 'These were not from the great forges of Yorkshire, or from the great steel works of Krupp [in Germany],' noted a friend. 'They were the work of an old, infirm man close on eighty years of age, who knew nothing about forging until after sixty, but who, when young, commenced by making everything he did as near perfect as it was possible, and who leaves as his monument the most perfect, the most novel forgings ever produced.'[3]

Even before his death in Monte Carlo in 1887, he gave away immense sums to educational and charitable causes in Manchester, including a lump sum of £100,000 to fund 30 technical scholarships a year. On his demise, a further £594,416 was given to his favourite causes, including Owens College, a forerunner of the University of Manchester, a colossal amount that equates to Bill Gates-style munificence in our times. To this day, graduates of the university collect their degree certificates in the Whitworth Hall. His legacy is one not merely of nuts and bolts, but also the Whitworth Art Museum and eponymous park in the centre of Manchester. He believed most profoundly in the value of technical education for working-men, and of the machinery that they went on to construct.

In 1979, the corroded wreck of a forgotten steamer, SS *Xantho*, was discovered off the cost of Western Australia, close to Port Gregory. The ship itself was not in the *Great Western* mode: there was nothing heroic or distinguished about the vessel. But there proved to be a bizarre double connection to Joseph Whitworth. The owner of the vessel was Charles Broadhurst, Whitworth's brother-in-law. He was the black sheep of the family, encouraged to seek his fortune in the furthest-flung corners of the empire, rather than in Lancashire. Charles sailed back from Australia for his sister's wedding to the great industrialist, no doubt entertaining and alarming fellow guests and family members with his madcap schemes for building a pearling empire in the South Pacific. On this trip home, he visited Glasgow and bought the *Xantho*, newly restored after 23 unexceptional years of service as a paddle-steamer in the waters off the Scottish coast.

The *Xantho* sailed to Australia and was briefly used to convey shell and aboriginal labourers between Australia and the trading centres of Batavia and Surabaya. In 1872, it hit a sandbank off Port Gregory and sank. Broadhurst went on to become a guano miner, his efforts to make a fortune in bird-dung as ill-fated as his pearling business, and the ship was forgotten. For more than a century the *Xantho* lay under water, the ship's machinery corroding in the seawater, accumulating a concrete-like crust of marine life. In 1983, marine archaeologists found that the ship's horizontal trunk engine, together with

---

3 John Fernie in *Proceedings of the Engineers' Club of Philadelphia* vol. 7, (1890), 31.

all its copper pipes, brasswork, lubricators, oil cups and other components, were intact. The wreck was raised in 1985, and over the course of the next quarter of a century, the engine was de-concreted, disassembled, cleaned-up and put back together and eventually displayed in Fremantle, at the Western Australian Museum's Shipwreck Gallery. The screws in the engine, boiler and pump were matched up with a screw pitch gauge, a handheld device like a penknife. Virtually all conformed to the British Standard Whitworth thread, thus demonstrating the global reach of Whitworth's manufacturing system.[4]

Whitworth's former homes are now both conference centres, and on the site of the shooting range is the Manchester University Botanical Garden. His firm was eventually merged with that of his great rival Sir William (subsequently Lord) Armstrong to create Armstrong Whitworth & Co, which was in time absorbed by Vickers.

* * *

In retirement, James Nasmyth extended his country house and renamed it Hammerfield. He planted trees, dabbled in photography and, from a study lined with pictures painted by his father and siblings, shelves laden with souvenirs of his spectacular career (including the slow-cooker he used to sustain himself in his time at Maudslay's), he engaged in correspondence with Herschel, Faraday and Samuel Smiles. In these bucolic surroundings, his mansion giving him splendid views of the ancient oaks of Penshurst Park and of the River Eden, he tinkered in his workshop and took an especial interest in astronomy and photography. He built three telescopes, two of the reflector sort (respectively, 13 and 20 inches in diameter) that he constructed himself, and a third refractor that was set up in an observatory in the garden, its walls housing a collection of fossils and hung with astronomical objects – for example, drawings 3-foot across of Jupiter's cloud belt and the Copernicus crater on the surface of the moon. Dangling from the ceiling was a mechanical contraption designed to show how large and small globes revolve around a centre of gravity. The 8-feet long telescope was set on a massive iron pillar to which it was attached by means of trunnions, more usually found controlling the elevation of cannons. With the delicacy of movement associated with the steam hammer, a little boy

---

4 The *Xantho* engine was made by John Penn of Greenwich in 1858, after the Crimean War was over. First intended for a Royal Navy gunboat, the order was subsequently cancelled and the engine was put into the *Xantho*. The ship was reconditioned in Glasgow and then sold to Broadhurst.

could direct this telescope across the sky with a touch of his finger. Country neighbours and eminent scientists were invited to come and gaze on the spectacular sight of the heavens yielded by this powerful device.[5]

Since the 1840s, Nasmyth had made a special study of the moon, making a series of detailed, 6-foot diameter drawings in black and white chalk on gray-tinted paper. 'I was thus able enabled to graphically represent the details with due fidelity as to form, as well as with regard to the striking effect of the original in its masses of light and shade.' He had shown some of these to Queen Victoria in 1851, and in retirement he took the work a stage further by making plaster models of his drawings, which he then photographed and turned into enormous transparencies. In 1874, he published *The Moon Considered as a Planet, a World and a Satellite* (co-written with James Carpenter), in which he voiced his theory that the craters of the moon were the result of volcanic activity. He also discovered the 'willow leaf' formations on the surface of the sun, a strange granulated pattern that is still difficult to observe. He gave an address to the Royal Astronomical Society on the temperature and condition of the planets Jupiter and Saturn. All this counts as a meaningful contribution to the science of astronomy, rather than the mere pottering of a rich amateur.[6]

Nasmyth had the rare pleasure of finding himself commemorated, while still only in his fifties, as the subject of one of Samuel Smiles's best-selling *Industrial Biographies*. 'You have quite immortalized me by not only saying so many handsome things about my doings,' he wrote to Smiles on the appearance of the book in November 1863, 'but by associating [me] so pleasantly with a set of men who deserved to be celebrated.' Until this point, Nasmyth explained to Smiles, biography had concentrated on military and naval men, but the achievements of engineers had the potential to be just as heroic. They deserved to be considered 'substantial benefactors of mankind,' he wrote. Smiles had originally contacted Nasmyth as a source for his account of Henry Maudslay, and Nasmyth had willingly penned a long memoir and answered numerous detailed queries about his mentor's life and personality. Evidently, Smiles decided that Nasmyth would be a good subject in his own right.

Samuel Smiles (1812–1904) is best known as the first industrial biographer and author of the immensely successful *Self-Help*. Smiles was at first a medical doctor, subsequently a popular lecturer and pamphleteer as well as businessman, in 1845 becoming secretary to the Leeds and Thirsk Railway.

---

5 'A Visit to the Observatory of the late James Nasmyth', in the *Western Daily Press*, 5 September 1890.

6 Pedro Ré, 'James Nasmyth's Telescopes,' on www.astrosurf.com/re

In 1854, he moved to London and for 12 years had the same position at the South Eastern Railway, which thus gave him some influence over Nasmyth's local station at Penshurst: the retired engineer wrote to Smiles in an attempt to stop the cutting down of some trees that beautified the approach to the station. Smiles's literary career began in earnest with the publication of his life of George Stephenson, published in 1857 and thus coinciding with the first year of Nasmyth's retirement. *Self-Help* had been written earlier but turned down by publishers; following the runaway success of the biography, this paean to Victorian individualism was brought out by John Murray in 1859 and achieved sales of 25,000 in the first year and 205,000 by 1905. It was a monumental bestseller, translated into many languages, bringing power and influence to its author and his ideas.

Whereas Samuel Johnson had written short lives of the poets, Samuel Smiles wrote about engineers, industrialists, ironworkers and toolmakers – and it was undoubtedly flattering for Nasmyth to be courted by Smiles as the latter began his research for his series of industrial biographies. Nasmyth, professing all the while to have had a sensation-free existence and to be devoid of literary skill, wrote scores of letters to Smiles, sketching forth details of his own career and memories of Maudslay in particular. When Maudslay and many of Smiles's other subjects were alive, the Victorian public did not yet exist and the British had not yet begun their love affair with their industrial entrepreneurs. The memorializing of their endeavours was a feature of a later stage of nineteenth-century civilization, when people had the time and luxury to look back on the achievements of those who had created the prosperity that they now enjoyed.

Despite his wealth, his scientific pursuits and glorification by Smiles, Nasmyth found retirement was not wholly tranquil. On 14 December 1864, eight years after Nasmyth left Manchester, T. S. Rowlandson, employee at Patricroft, delivered a public lecture in the vicinity of the factory confirming what so many in Manchester must have known already: that Nasmyth was only in a tendentious way the inventor of the steam hammer.

According to Nasmyth's own account, he first had the idea for the hammer in November 1839 when the directors of the Great Western Steam Ship Company asked him for help in building the engines for the SS *Great Britain*, the second of Isambard Kingdom Brunel's great ships, originally conceived (like most early steam-powered vessels) as a hybrid between a sailing ship and a paddle steamer. This would have required a massive axle to run across the ship from the engine to drive the paddle wheels. Forging and shaping such an enormous item, with a 30-inch diameter shaft, was beyond the scope of existing tools, and thus a new invention was called for. 'In little more than half an hour,' Nasmyth claimed, 'I had the whole contrivance in all its executant details, before me in a page of my Scheme Book'.[7] Brunel did not in the end commission the design, as he switched from paddle technology to the newly invented screw propeller, and so the drawing stayed in Nasmyth's notebook.

Nasmyth subsequently wrote to all the great forge masters, including Acramans and Morgan of Bristol, Benjamin Hick, Rushton and Eckersley of Bolton and Howard and Ravenhill of Rotherhithe, in an attempt to sell the product, but found no interest. Later that year, when Nasmyth was absent, Eugene Schneider, proprietor of the giant Le Creusot ironworks in France, visited Patricroft, together with his chief engineer, Francois Bourdon. The French were customers as well as putative competitors, having travelled to Manchester to buy machine tools, and Nasmyth's business partner Gaskell showed them around, calculating that the risk of having the firm's intellectual property stolen was outweighed by the prospect of getting orders. Gaskell showed the visitors the Scheme Book with the historic drawing of the steam hammer. Nasmyth knew nothing about the incident until April 1842, when he visited the Le Creusot works and was confronted with his own invention, busy producing enormous cranks for marine engines, precisely the kind of large-scale work that no one could as yet make at home in Britain.

'How did you forge that shaft?' Nasmyth asked.

'Why with your hammer, to be sure,' was allegedly the Frenchman's reply.

Great, indeed, was Mr Nasmyth's surprise, *wrote Smiles*, for he had never yet seen the hammer, except in his own drawing! [The Frenchman] said he had been so much struck with the ingenuity and simplicity of the arrangement [when shown it in the scheme book], that he had no sooner returned than he set to work, and had a hammer made in general accordance with the design Mr. Gaskell had shown him, and that its performance had answered his every expectation.[8]

For all the complicated self-justifications that ensued, Nasmyth had clearly not appreciated the machine's commercial potential. A bitter and protracted dispute broke out between Nasmyth and the French firm. Le Creusot claimed that they had developed their own steam hammer before they had seen Nasmyth's drawing, which, they said, was too rudimentary to have been of any practical use. Nasmyth

7 Nasmyth, *Autobiography*, 231.

8 Smiles, *Industrial Biography*, 287–88.

claimed that, on the contrary, the machine he had seen in France lacked some of the utility of his own invention. He maintained that in his 'rude and hasty first sketch' he had 'hit the mark so exactly, not only in the general structure but in the details'. According to John Cantrell, the leading Nasmyth scholar, this claim is 'both exaggerated and misleading'.[9] Nasmyth did not at first come up with anything more than the crude conception of the design, and certainly not the self-acting refinements that made the tool so universally useful. The page full of drawings was published for all to see in his *Autobiography*. The sketch that purports to show the tool halfway towards its ultimate state of automated glory, was almost certainly added long after the event (i.e., in 1842, when Nasmyth returned from France and hurriedly produced the patent specification). Le Creusot carried on making enormous steam hammers in their own name.

By the time of the 1864 lecture, Robert Wilson had become head of the firm that still carried James Nasmyth's name. He had taken out many patents in his own right and was clearly an engineer of great distinction. Perhaps he was attacking his former master and fellow Scot in order to settle old scores? He was certainly responding to the claims in Smiles's *Industrial Biography: Ironworkers and Toolmakers* (1863) that uncritically identified Nasmyth as the sole inventor. Those claims were given credence by no less a figure than William Fairbairn, who stated that Nasmyth, with the mere *assistance* of Mr. Wilson, had introduced the new and simple device for working the hammer (i.e., the self-acting device). Disobligingly, from Nasmyth's point of view, Holbrook Gaskell came forward to confirm the truth of Wilson's version of the story:

That to Mr. Nasmyth is due the original conception of the direct action steam hammer, I have frequently testified, and am prepared to maintain: but that either he or anyone else had any conception of the great future which awaited his invention, I distinctly deny. [In its original form] the utility of the machine was extremely restricted [so] I felt desirous that it should be made self-acting, so that it might be worked at a higher speed, and thereby be adapted to all ranges of forgings[; ...] the result was [Wilson's] very beautiful invention of the self-acting motion.[10]

---

9 See J. A. Cantrell, *James Nasmyth and the Bridgewater Foundry* (1984), Chapter 6, for a full account of the competing claims for the genesis of the steam hammer. The French have a valid claim to have built the first self-acting steam hammer, but the Le Creusot models were influenced by Nasmyth's designs.

10 Letter to Wilson, 11 November 1864, cited in T. S. Rowlandson's *History of the Steam Hammer*, 7.

FIGURE 8 A steam hammer at work, oil painting from 1871 (Courtesy of Science Museum, London)

Then Fairbairn himself delivered the coup de grace, explaining that he had made his statement 'on the authority of Mr. Nasmyth himself'. He added, significantly: 'considering Mr. Nasmyth's high position as a gentleman and a man of science, I never doubted the information I received to be otherwise than in strict accordance with the truth' (Figure 8).[11]

How should we interpret this episode? Was Nasmyth an unscrupulous scoundrel, who stole the work of others, or was he guilty of no more than what the Germans call a *Kavaliersdelikt*, an insignificant infraction of an ethical code not applicable to a man of his stature? Nasmyth was technically the inventor, and was legally within his rights to appropriate the intellectual property of refinements designed by a mere employee; this, after all, was how Charles Babbage had treated Joseph Clement. But, in 1843 Nasmyth had taken out a patent for the self-acting motion of the hammer *in his own name only*, and subsequently concealed this from both Wilson and Gaskell. It was the

11 Letter from Fairbairn to Rowlandson, 29 December 1865, cited in the *History*, 23.

discovery of this duplicity, many years later, as much as the one-sided account in Smiles's biography, that encouraged Wilson to host the lecture, to speak out and publicize the truth. There was extensive press coverage following the lecture, and most gave the benefit of the doubt to Wilson. 'It was unfortunate for Mr. Wilson's pocket and fame that [at the time of the invention] he was then a working man,' opined the *Preston Chronicle* on 18 March. 'Had he been what he is now, or had he had friends, it is scarcely possible that we should have heard the name of Nasmyth in connection with the steam hammer.' Nasmyth's standing as a gentleman, as well as engineering pioneer, was under attack. One has to wonder whether this public assault on his integrity was one reason why he was never honoured with a title, unlike Fairbairn and Whitworth, who were made baronets on the same day in October 1869, along with Titus Salt, the enlightened proprietor of the Saltaire mill and model village in Bradford, Yorkshire.

Nasmyth did not deign to respond. 'The best way to meet such detractors is silence,' he told Smiles, privately. 'All who have invented machines have had to withstand the stings of "wasps and mosquitoes." ' He denounced Wilson's 'systematic, grasping conceit'.[12] Not only had he renamed his mansion, his coat of arms featured a hammer, an impression of which was to be found on every letter that left his desk. His whole identity was bound up with the steam hammer, and under no circumstances was he going to change his story. He paid Smiles the princely sum of £1,000 to ghost write his *Autobiography*, published in 1883. This could have been an opportunity to set the record straight and provide at least acknowledgment of Wilson's role. But the book did nothing of the kind: it was a deft propaganda blow struck with the precision and force of a self-acting steam hammer, ensuring that the history of nineteenth century engineering was told on his terms.

Amid his extensive correspondence with his biographer and distinguished men of science, Nasmyth also kept up with the inaptly named Virtue Squibb, whom he called Floss:

My Dear Floss,

I hope this shabby little Billy finds you and Minnie in prime condition as well as all dear to you – I look forward to the happiness of seeing you and having a gossip when I run in this Evening say about 7. And if you are so disposed we can go and have a mild little Blow out at the usual place about

---

12 Letter, May 1881.

9. After which I must see you safe into a cab and say good night and all
happiness attend you till the next time.

Yours ever affectionately
The old file
His + marke

The mark was the thumbprint with which Nasmyth signed off the
Autobiography and his letters to scientific worthies; the 'old file', an allusion to
the early days at Maudslay when he used to wield a file.

My Dear Floss,

I shall be in town for a few hours tomorrow [. ...] I have to return by the
Five ock train home, but as we shall have a good two hours at the grill room
tables we can try to make the most of it.

You will find a pudgey old party parading about the door at Jermin St
usual place awaiting you in the hope of finding you in prime condition.

Yours Ever faithfully,
The old File
Hys + Marke

These letters show Nasmyth as a sentimental father and adulterer, aspects of
his personality that were not celebrated in the *Autobiography*.[13] Floss lived at 28,
Lupus Street, Pimlico, and became his mistress at some point in the 1850s,
and Minnie, born in 1858 or 1859, was their daughter. He was in the habit of
taking the train up from Kent and meeting Virtue at the Criterion Restaurant
in Piccadilly. On Virtue's death in 1885, Nasmyth wrote to Minnie with a
post office order for £10 and the promise of a further £1,000 'to be sacredly
kept by you for your own use as a sacred remembrance of your dear mother'.
Despite the engineer's genuine fondness to his daughter, it somehow does
not come as a surprise to find that she never received the £1,000. Nasmyth

---

13 These letters were unearthed by the indefatigable Dr Cantrell and are kept at the
National Library of Scotland. The catalogue entry reads: 'Two letters, 1880, n.d., of
James Nasmyth, engineer, to his mistress Virtue Squibb ('Emily Russell'), together with
one letter of Virtue Squibb to their daughter Minnie, and one of Nasmyth to Minnie,
1885. The letters are accompanied by notes compiled by Dr John Cantrell, Stockport,
Cheshire.'

himself died on May 7, 1890, at Bailey's Hotel in Gloucester Road, South Kensington, in his 82nd year. After cremation, his remains were packed up and sent to Scotland by train, wrapped in a brown paper bag, thus attracting a lower fare than the full passenger rate normally charged for transporting human ashes. Officially childless, he left the bulk of his enormous fortune (still worth £250,000 on his death) to his widow and various charities, including a still-disbursing fund, named after his father, Alexander, for impoverished Scottish artists.

After Nasmyth died, his country house was taken on by Lord Ronald Sutherland-Leveson-Gower, celebrated Uranist, friend of Oscar Wilde, and probable model for Lord Henry Wotton in *The Picture of Dorian Gray*. The company that bore Nasmyth's name was developed into one of the world's leading specialist locomotive engineering operations, producing over the course of precisely 100 years 1,652 locomotives for countries as various as Thailand, India, Malaysia and Sweden. Though the firm finally closed down in 1938, some of James Nasmyth's early customers survive to this day: De La Rue, for example, is an independent public company, producing banknotes, while Tootal is still in the textiles business, producing coveted silk ties and cravats.

* * *

There are other corporate survivors from the Maudslay era: more than a century and a half after the Great Exhibition, Chubb is still a leading lock manufacturer, but part of a Swedish conglomerate. Messrs Bramah are still proudly British, and the company maintains a showroom in London's West End. The lock built by Maudslay and picked by Hobbs is on display at the Science Museum, an institution established with the profits of the Great Exhibition. The business established by Henry Maudslay's friend, Bryan Donkin, is still trading under the founder's name, a world leader in gas-control equipment.

Alas, Maudslay's itself has not survived. It was thriving when Joshua Field died in 1863, heaped with honour and distinctions, and successive generations of Maudslays kept the firm going. But by the 1890s, one senses that their heart was not in it. Their descendant, Richard Maudslay, speculates that by this time, many family members were living off the dividends paid by the company, and that the entrepreneurial verve of the early generations was extinguished. Alfred Maudslay (1850–1931), a great-grandson, became a notable explorer of Central America, finding the excavation of lost Mayan cities in the Guatemalan jungle to be more congenial than fighting to protect the firm's position in markets increasingly open to international competition. Loss-making throughout the 1890s, the firm was declared 'hopelessly insolvent' in October 1899. Other descendants maintained the family engineering tradition: Cyril Maudslay (1875–1962), a great-grandson of the founder, established the Maudslay Motor Company in 1902, and this Coventry-based company remained a leading maker of cars and lorries until the 1940s. His brother, Athol, was a well-known amateur inventor, while still another brother, Herbert, was the inventor of the spinnaker sail, named after his yacht the *Sphinx* (sphinx acre), and beloved of adventurous sailors. Richard, in the current generation, was a main board director of Rolls-Royce with responsibility for the engineering company's non-aero businesses, including his ancestor's specialisms of power generation and marine propulsion. As for the remains of Henry Maudslay, himself, Greenwich Council cleared the graveyard in which he was buried in 1966, and his tomb was destroyed. Showing a sadly predictable sensitivity to popular culture, the only grave preserved was that of Tom Cribb, a prize-fighter. Remnants of Maudslay's tomb are kept in the basement of a museum in Greenwich.

* * *

In the 185 years since his death, Maudslay's reputation among all but ultra-specialists has been subject to similar neglect. The Portsmouth block factory is all but derelict and closed to the public, despite its importance to the history of industry and naval warfare. Some of the Portsmouth machines sit in the Science Museum, as beautiful and intricate as sculptures, but with none of the power, movement and purpose that excited contemporaries. As L. T. C. Rolt discerned, the revolution that he inspired took place on the unglamorous shop floor, in workshops and factories, and its output was less obviously spectacular than the railways, tunnels, bridges and other triumphs of Victorian civil engineering, which of course could not have come into being without the influence of Maudslay and his circle.

Supremely talented though he was, Maudslay's inventions were not effusions of personal genius, but sprang from multiple, eclectic influences, rooted in the craft economy of the late eighteenth century. In his own person, he assimilated diverse artisanal skills, from blacksmithery to instrument making, building with his own hands machines that operated to unprecedented standards of accuracy and strength. He came to prominence in the service of Britain's war effort, and although the naval and military context of his early life at Woolwich and Portsmouth explains how he learnt his trade and obtained many of his commissions, his chief impact was on the peaceful productive capacity of the nation. As Karl Marx recognized, Maudslay's self-acting slide-rest was the first crucial step towards the full panoply of mechanization on display at the Great Exhibition, begetting other machines that begat others that led, like a great chain of mechanical being, to the automation and industrialization of our own times. Maudslay was the bridge between the age of the craftsman and that of the machine: an artisan who transcended the limitations of mere human skill to create machines that could take advantage of steam power and harness the potential of iron and steel and dispense with wood and brass.

Roberts, Nasmyth, Whitworth and others such as Clement and Muir learned their craft from direct observation of Maudslay at work in his factory in the Westminster Bridge Road. Despite their different backgrounds, temperaments and ultimate success in the world, they all started with talent and little by way of material resources, some making the tools they would use to build their businesses. It took this generation, those I have dubbed the Maudslay men, to disseminate the principles and practices of the master and thus follow mechanization through to its logical conclusion, where the Iron Men they constructed could do the job of flesh and blood human beings, creating mass production and interchangeable components, introducing standardization and unprecedented levels of practical accuracy, building machines to build other machines, and in the process transforming the cotton industry, enabling the railways to be developed, and promoting the explosive expansion of the engineering industry more generally. Likewise, the skill of Joshua Field and Joseph Maudslay in particular helped power the shipping industry.

E. J. Hobsbawm wrote that the era covered by this book saw 'the most fundamental transformation of human life in the history of the world recorded in written documents'.[14] In the half century leading up to the Great Exhibition, Britain became 'the undisputed economic leader of the world, enjoying a newly found political prestige and hegemonic power'.[15] This economic power buttressed the empire and undergirded the Victorian era, the true Golden Age of Great Britain. The social and political turbulence of the early part of the century gave way to the so-called Age of Equipoise. Thomas Carlyle, Charles Dickens and Karl Marx saw the associated mechanization as dehumanizing and spiritually impoverishing. Others, including Joseph Whitworth, believed

---

14 This is the opening sentence of Eric Hobsbawm's magisterial *Industry and Empire: The Birth of the Industrial Revolution* (1969).

15 Mokyr, *The Enlightened Economy*, 3.

that the kind of self-acting or semi-automatic machines on display in Hyde Park liberated the masses from the drudgery of manual work and created better-paid, more highly skilled jobs, for more people. 'I feel some complacent pride in having done some good work in contriving tools now universally employed which have had no small influence in doing away with the dependence on mere brute force,' wrote Nasmyth. 'Every workshop in the world where machinery is made bears witness to this.'[16] Or, as J. S. Mill put it, 'human beings are no longer born to their place in life [...] but are free to employ their faculties and such favourable chances on offer, to achieve the lot which may appear to them most desirable'.[17]

The debate among historians still rages. But for all the poverty and squalor associated with rapid industrialization, the expanding population enjoyed enduring improvements in living standards, and the economy began to grow at an unprecedented rate. In the long run, writes Robert C. Allen, this growth 'compounded to the mass prosperity of today'.[18] Deirdre McCloskey has argued that the period should on this account be described as the 'Great Enrichment', rather than the Industrial Revolution. 'The Great Enrichment is the rise of real income per head since 1800, 30 times – that's 2900%, class,' she tweeted on 3 June 2013. 'Investing and exploiting can't cause.'[19] Queen Victoria, and the millions of others who attended the Great Exhibition, would not recognize the language of modern economic historians, but they would understand the sentiment.

Maudslay's immense contribution to this process deserves to be properly acknowledged. In the words of James Nasmyth, an unsympathetic man but a true disciple, Henry Maudslay led 'a useful life [...] enthusiastically devoted to the grand object of improving our means of producing perfect workmanship and machinery [and was distinguished by] the indefatigable care which he took in inculcating and diffusing among his workmen, and mechanical men generally, sound ideas of practical knowledge, and refined views of construction'.[20] Quite literally, Maudslay and his men helped shape the modern world.

---

16 Letter to Smiles, 7 July 1862.

17 J. S. Mill, *The Subjection of Women* (1869), Chapter 1.

18 Robert C. Allen, *Global Economic History: A Very Short Introduction* (2009), 27.

19 See http://www.deirdremccloskey.com/ for further trenchant argument along these lines.

20 Cited in Smiles, *Industrial Biography*, Chapter 12.

ACKNOWLEDGEMENTS

This book could not have been written without the support of my wife, Jane, and children, Max, Pippa and Munro, who for a period of years have been dragged along to see steam engines, mills and museums, and as a result know more about the Industrial Revolution than they may have wished to. Thanks to Richard Maudslay, CBE, a descendant of this book's central figure and a distinguished engineer in his own right, who took an early interest in this project and has provided helpful insights throughout. Ben Russell, curator of mechanical engineering at the Science Museum, London, read an early draft and showed me round the storerooms of the museum, where many exotic and fascinating machines are hidden away from public view. The manuscript was also read by John Ditchfield, engineer and contributor to the excellent Grace's Guide to British Industrial History website (www.gracesguide.co.uk), who very generously gave of his time and specialist knowledge to make detailed comments, while John Cantrell, the leading James Nasmyth scholar, took the time to read a number of chapters in draft. Thanks to Edgar Feuchtwanger and Jim Ring, who read early versions, and also to George Blumberg, who shared his enthusiasm for machine tools on an enjoyable walk on the South Downs. John Hatch kindly showed me the beam engines in situ at the Ram Brewery in Wandsworth. I am indebted to all the above for generously sharing of time and knowledge, but of course I alone take responsibility for any errors of fact or judgement. Thanks to the staff of the London Library and of the British Library, to Karyn Stuckey and Adrian Clement of the Institution of Mechanical Engineering, to Carole Morgan of the Institution of Civil Engineers and to Jacqueline Apperley of George Stephenson's birthplace at Wylam, Northumberland. Finally, thanks to Tej Sood, publisher, for his enthusiastic support for this book, and to the team at Anthem Press for seeing *Iron Men* through to publication.

BIBLIOGRAPHY

Allen, Robert C. *The British Industrial Revolution in Global Perspective.* Cambridge: Cambridge University Press (2009).

———. *Global Economic History: A Very Short Introduction.* Oxford: Oxford University Press(2011).

Anon. *The Bramah Lock Controversy: Extracts from the Press.* London: T. Brettel (1851).

Armstrong, John, and David M. Williams. 'The Beginnings of a New Technology: The Constructors of Early Steamboats 1812–22', *International Journal for the History of Engineering and Technology*, vol. 81, no. 1 (January 2011): 1–21.

Atkinson, Norman. *Sir Joseph Whitworth: The World's Best Mechanician.* Gloucester, UK: Sutton (1996).

Auerbach, Jeffrey A. *The Great Exhibition of 1851: A Nation on Display.* New Haven: Yale University Press (1999).

Babbage, Charles. *On the Economy of Machinery and Manufactures.* Cirencester, UK: Echo Library (1846).

———. *Passages from the Life of a Philosopher.* New Jersey: Rutgers University Press (1864).

Bailey, Michael, ed. *Robert Stephenson: Eminent Engineer.* London and New York: Routledge (2003).

Beamish, Richard. *Memoir of Sir Marc Isambard Brunel.* London: Longman, Green, Longman and Roberts (1862).

Bentham, Maria Sophia. *The Life of Brigadier-General Sir Samuel Bentham, KSG.* London : Longman, Green, Longman and Roberts (1862).

Berg, Maxine. *The Machinery Question and the Making of Political Economy 1815–1848.* Cambridge: Cambridge University Press (1980).

———. *Age of Manufactures, 1700–1820: Industry, Innovation and Work in Britain.* London and New York: Routledge (1994).

Buchanan, Robert Angus. *The Engineers: a History of the Engineering Profession in Britain, 1750–1914.* London: Jessica Kingsley Publishers (1989).

———. *Brunel: The Life and Times of Isambard Kingdom Brunel.* London and New York: Hambledon and London (2002).

———. 'Gentlemen Engineers', *Victorian Studies*, vol. 26 (1983): 407–29.

———. 'Science and Engineering: A Case Study in British Experience in the Mid-Nineteenth Century', *Notes and Records of the Royal Society of London*, vol. 32, no. 2 (March 1978): 215–23.

Buchanan, Robertson. *Practical Essays on Mill Work and Other Machinery.* London: J. Weale (1841).

Burton, Anthony. *Richard Trevithick: Giant of Steam.* London: Aurum Press Ltd (2000).

Buxton, L. H. Dudley. 'Charles Babbage and His Difference Engines', *Transactions of the Newcomen Society (TNS)*, vol. 14 (1934): 43–65.

Cantrell, John. *Nasmyth, Wilson & Co: Patricroft Locomotive Builders.* Stroud: History Press Ltd. (2005).

———. 'James Nasmyth and the Steam Hammer', *TNS*, vol. 56 (1984): 133–38.

———. *James Nasmyth and the Bridgewater Foundry: A Study of Entrepreneurship in the Early Engineering Industry.* Dover: Manchester University Press (1984).

———. 'Two Maudslay Protégés: Francis Lewis and George Nasmyth' *TNS*, vol. 73 (2003): 257–74.

Cantrell, John, and Gillian Cookson, eds. *Henry Maudslay and the Pioneers of the Machine Age.* Stroud: NPI Media Group (2002).

Carlyle, Thomas. 'Signs of the Times', *Edinburgh Review* (1829).

Carlyle, Thomas. *Selected Writings.* London: Penguin (1980).

Catterall, G. S. *The Life and Works of Richard Roberts with Special Reference to the Development of the Self Actor Mule.* MSc. thesis, University of Manchester, 1975.

Coad, Jonathan. 'Chatham Ropeyard', *Post Medieval Archaeology*, vol. 3 (1969).

———. *The Portsmouth Block Mills: Bentham, Brunel and the Start of the Royal Navy's Industrial Revolution.* Swindon: English Heritage (2005).

Chaloner, W. H. 'New Light on Richard Roberts, Textile Engineer (1789–1864)', *TNS*, vol. 41 (1968–69).

Chapman, S. D. *The Cotton Industry in the Industrial Revolution.* London: Macmillan (1972).

Chrimes, Michael, Julia Elton, John May and Timothy Millet. *The Triumphant Bore: A Celebration of the Thames Tunnel.* London: Thomas Telford (1993).

Claxton, Christopher. *The Logs of the First Voyage, made with the unceasing aid of steam, between England and America . . . by the Great Western.* Bristol: Mirror Office (1838).

Cookson, Gillian. 'Innovation, Diffusion, and Mechanical Engineers in Britain, 1750–1850', *Economic History Review*, vol. 47, no. 4 (1994): 749–53.

Cooper, Carolyn. 'The Portsmouth System of Manufacture', *Technology and Culture*, vol. 25, no. 2 (April 1984): 182–225.

Davenport-Hines, Richard. *Capital, Entrepreneurs and Profits.* Abingdon: Routledge (1990).

Davis, John. *The Great Exhibition.* Stroud: Sutton Pub Ltd (1999).

Deane, Phyllis. *The First Industrial Revolution.* Cambridge: Cambridge University Press (1965).

Dickinson, H. W. 'Richard Roberts, His Life and Inventions', *TNS*, vol. 25 (1945–47).

Dickinson, H. W. 'The Taylors of Southampton: Their Ships' Blocks, Circular Saw, and Ships' Pumps'. *TNS*, vol. 29, no. 1 (1953): 169–78.

Evans, F. T. 'The Maudslay Touch: Henry Maudslay, Product of the Past and Maker of the Future.' *TNS*, vol. 66 (1994): 153–74.

Elliot, W. Hume. *The Story of the 'Cheeryble' Grants.* Manchester: Hard Press Publishing (1906).

Emerson, Ralph Waldo. *English Traits.* London: G. Routledge & Co. (1856).

Engels, Friedrich. *The Condition of the Working Class in England in 1844.* Leipzig: Otto Wigand (1845).

Fairbairn, William. *An Account of the Construction of the Britannia and Conway Tubular Bridges.* London: John Weale and Longman, Brown, Green and Longmans (1849).

———. *On Tubular Girder Bridges.* London: W. Clowes and Sons (1851).

———. *Treatise on Mills and Millwork*, 2 vols. (Vol. 1, London: Longman, Green, Longman and Roberts (1861) and Vol. 2: London: Longman, Green, Longman, Roberts and Green (1863)).

———. *The Life of Sir William Fairbairn, Bart.* ed. and completed William Pole, first published London: Longmans, Green (1877).

Farey, John. *A Treatise on the Steam Engine.* London: Longman, Rees, Orme, Brown and Green (1827).

Faucher, Leon. *Manchester in 1844: Its present condition and future prospects; translated from the French by a member of the Manchester Athenaeum.* London: Cass (1844).

Field, Joshua. 'Diary of a Tour in 1821 through the Midlands', *TNS*, vol. 6 (1925): 1–41.

———. 'Diary of a Tour in 1821 through the Provinces', *TNS*, vol. 13 (1932): 15–50.

Francis, J. A. *A History of the English Railway: Its Social Relations and Revelations 1820–1845.* London: Longman, Brown, Green and Longmans(1851).

Freeman, Michael J. *Railways and the Victorian Imagination.* New Haven: Yale Univeristy Press (1999).

Garfield, Simon. *The Last Journey of William Huskisson.* London: Faber & Faber (2002).

Gaskell, Elizabeth. *Mary Barton.* London: Chapman & Hall (1848).

———. *North and South.* London: Chapman & Hall (1854–1855).

Gaskell, Peter. *The Manufacturing Population of England, its Moral, Social and Physical Condition.* London: Baldwin and Cradock (1833).

Gilbert, K.R. *The Portsmouth Blockmaking Machinery.* Chicago: University of Chicago Press (1964).

———. *Henry Maudslay: Machine Builder.* London: HMSO (1971).

Goodeve, Thomas Minchin and Charles Percy Bysshe Shelley. *The Whitworth Measuring Machine.* London: Longmans, Green, and Co. (1877).

Gordon, Robert J. 'Is Economic Growth Over? Faltering innovation confronts the six headwinds', *Centre for Economic Policy Research, Policy Insight No. 63* (September 2012).

Griffin, Emma. *Liberty's Dawn: A People's History of the Industrial Revolution.* New Haven and London: Yale University Press (2013).

Griffiths, Denis. *Brunel's Great Western.* Wellingborough: Patrick Stephens (1985).

Griffiths, John. *The Third Man: The Life and Times of William Murdoch, Inventor of Gas Lighting.* London: Andre Deutsch Ltd (1992).

Habakkuk, H. J. *American and British Technology in the Nineteenth Century.* Cambridge: Cambridge University Press (1967).

Henderson, W. O. *J. C. Fischer and His Diary of Industrial England, 1814–1851.* New York: A. M. Kelley (1966).

Hills, The Rev Dr. Richard. *The Life and Inventions of Richard Roberts, 1789–1864.* Ashbourne: Landmark (2002).

Hobbs, Alfred Charles. *Rudimentary Treatise on the Construction of Locks.* London: John Weale (1853).

Hobsbawm, Eric. *Industry and Empire: The Birth of the Industrial Revolution.* London: Penguin Books (1969).

Holtzapffel, Charles. *Turning and Mechanical Manipulation II.* London: Holtzapffel. (1850).

Hounshell, David. *From the American System to Mass Production, 1800–1932.* Baltimore: Johns Hopkins University Press (1985).

Hughes, Thomas. *Account of the Lock-out of Engineers, 1851–52, prepared for the National Association for the Promotion of Social Science.* Cambridge and London: Macmillan (1860).

Jeaffreson, John Cordy and William Pole, *Life of Robert Stephenson.* London: Longmans (1864)]

Jennings, Humphrey. *Pandaemonium 1660–1886: The Coming of the Machine Age as Seen by Contemporary Observers.* London: André Deutsch (1985).

Kargon R. H. *Science in Victorian Manchester: Energy and Expertise.* Manchester: Manchester University Press (1977).

Kay-Shuttleworth, Sir James. *The Moral and Physical Condition of the Working Classes Employed in the Cotton Manufacture in Manchester.* London: Ridgway (1832).

Knight, Roger. *Britain Against Napoleon: The Organisation of Victory.* London: Penguin (2013).

Landes, D. S. *The Unbound Prometheus.* Cambridge: Cambridge University Press (1969).

Lardner, Dionysus, Essay on Babbage in *Edinburgh Review* no. 120 (July 1834): 263–327.

———. *Railway Economy: A Treatise on the New Art of Transport.* New York: Harper & Brothers (1850).

———. *The Steam Engine: Steam Navigation, Roads and Railways, Explained and Illustrated.* London: Taylor Walton Maberly (1851).

Lea, F. C. *Sir Joseph Whitworth: A Pioneer of Mechanical Engineering.* London: British Council by Longmans (1946).

Lee, Charles E. 'Railway Engineering: Its Impact on Civilisation', *TNS*, vol. 36 (1963) 109–35.

Marcus, Steven. *Engels, Manchester and the Working Class.* New York: Random House (1974).

Marshall, Geoff. *London's Industrial Heritage.* Stroud: The History Press (2013).

Marx, Karl. *Capital.* First published in three volumes: 1867, 1885 and 1894,

———. *The Poverty of Philosophy.* With an introduction by Friedrich Engels. Edited by C. P. Dutt and V. Chattopadhyaya. London: Martin Lawrence (1936).

Maudslay, Cyril, and J. Foster Petree, *Henry Maudslay, 1771–1831, and Maudslay, Sons & Field, Ltd.,* a memoir prepared for private circulation (1948).

Mercer, Stanley. 'Trevithick and the Merthyr Tramroad', *TNS*, vol. 26 (1947–49): 89–103.

Mill, John Stuart. *Autobiography.* London: Longmans (1873).

Mokyr, Joel. *The Enlightened Economy: Britain and the Industrial Revolution 1700–1850.* New Haven and London: Yale University Press (2009).

Montefiore, Simon. 'Prince Potemkin and the Benthams', *History Today*, vol. 53, no. 8 (August 2003).

Musson A. E. and Robinson, Eric. *Science and Technology in the Industrial Revolution.* Toronto: University of Toronto Press (1969).

Musson A. E. 'Joseph Whitworth: Toolmaker and Manufacturer', *Engineering Heritage: Highlights from the History of Mechanical Engineering*, 1 (1963): 124–29.

———. 'Joseph Whitworth and the Growth of Mass Production Engineering', *Business History* no. 17 (1975): 109–49.

Odlyzko Andrew. 'Crushing national debts, economic revolutions, and extraordinary popular delusions,' University of Minnessota Working Paper 2012, accessed on http://www.dtc.umn.edu/~odlyzko/doc/mania05.pdf on 27 June 2015.

Petree, J. Foster, 'Maudslay Sons & Field as General Engineers', *TNS*, January 1934, 39–61.

Phillips, Sir Richard. *A Morning's Walk from London to Kew*. London: J. Souter (1817).

Riello, Giorgio. *Cotton: The Fabric that Made the Modern World*. Cambridge: Cambridge University Press (2013).

Riley, Ray, ed. Transactions of the Naval Dockyards Society: Transactions of the Ninth Annual Conference held at Portsmouth April–May 2005: Portsmouth Dockyard in the Age of Nelson, vol. 1 (July 2006).

Roe, Joseph Wickham. *English and American Tool Builders*. New Haven: Yale University Press (1916).

———. 'Interchangeable Manufacture', *TNS*, vol. 17 (1936–37).

Rolt, L. T. C. *Tools for the Job: A Short History of Machine Tools*. London: Batsford (1965).

———. *Victorian Engineering*. London: Allen Lane (1970).

Rosenberg, Nathan, ed. *The American System of Manufactures: The Report of the Committee on the Machinery of the United States 1855, and the special reports of George Wallis and Joseph Whitworth*, with an introduction by Nathan Rosenberg. Edinburgh: Edinburgh University Press (1969).

Ross, David. *George & Robert Stephenson: A Passion for Success*. Stroud: The History Press Ltd. (2010).

Rowlandson, T. S. *History of the Steam Hammer*. Eccles: A. Shuttleworth (1864).

Russell, Ben. *James Watt: Making the World Anew*. (2014).

Schaffer, Simon. 'Babbage's Intelligence: Calculating Engines and the Factory System', *Critical Inquiry* 21 (Autumn 1994): 203–27.

Schivelbusch, Wolfgang. *The Railway Journey: The Industrialization of Time and Space in the 19th Century*. Berkeley: University of California Press (1986).

Secord, James A. *Visions of Science: Books and Readers in the Victorian Age*. Oxford: Oxford University Press (2014).

Semmens, P. W. B. and A. D. Goldfinch. *How Steam Engines Really Work*. Oxford: Oxford University Press (2000).

Simmons, Jack. *The Victorian Railway*. New York: Thames and Hudson (1991).

Smiles, Robert. *Brief Memoir of the Late William Muir*. Published circa 1888, accessible on www.gracesguide.co.uk.

Smiles, Samuel. *Life of George Stephenson*. London: J. Murray (1857).

———. *Self-Help*. London: John Murray (1859).

———. *Lives of the Engineers*. 5 vols. London: John Murray (1862).

———. *Industrial Biography: Ironworkers and Toolmakers*. Boston: Ticknor and Fields (1863).

———. ed. *James Nasmyth. An Autobiography*. London: J. Murray (1883).

Sussman, Herbert. *Victorian Technology: Invention, Innovation and the Rise of the Machine*. Santa Barbara: Praeger (2009).

Swade, Doron. *The Cogwheel Brain: Charles Babbage and the Quest to Build the First Computer*. London: Little Brown (2000).

Tennent, Sir James Emerson. *The Story of the Guns*. London : Longman, Green, Longman, Roberts, & Green (1864).

von Tunzelmann, G. N. *Steam Power and British Industrialisation to 1860*. Oxford: Oxford University Press (1978).

Ure, Andrew. *A Dictionary of Arts, Manufactures and Mines*. New York: D. Appleton & Co (1842).

———. *The Philosophy of Manufactures*. 3rd ed. London: Charles Knight (1861).

———. *The Cotton Manufacture of Great Britain* in 2 vols. London: H. G. Bohn (1861).

Walling, R. J. ed. *The Diaries of John Bright*. New York: William Morrow (1930).

Walmsley, Robert. *Peterloo: The Case Re-opened*. Manchester: Manchester University Press (1969).

Weale, John. *Quarterly Papers on Engineering*, vol. 6. London (1846).

Weightman, Gavin. *The Industrial Revolutionaries: The Creation of the Modern World 1776–1914*. London: Atlantic Books (2007).

Whitworth, Joseph. *Miscellaneous Papers on Mechanical Subjects*. London: Longman, Brown, Green, Longmans, and Roberts (1858).

Wolmar, Christian. *Fire and Steam: How the Railways Transformed Britain*. London: Atlantic Books (2007).

INDEX